Lecture Notes in Mathematics

Edited by A. Dold, Heidelberg and B. Eckmann, Zürich

368

R. James Milgram

Stanford University, Stanford, CA/USA

Unstable Homotopy from the Stable Point of View

Springer-Verlag
Berlin · Heidelberg · New York 1974

AMS Subject Classifications (1970): 55-02, 55E10, 55E20, 55E25, 55E40, 55E99, 55D20, 55D35, 55D40, 55D99

ISBN 3-540-06655-1 Springer-Verlag Berlin · Heidelberg · New York
ISBN 0-387-06655-1 Springer-Verlag New York · Heidelberg · Berlin

Offsetdruck: Julius Beltz, Hemsbach/Bergstr.

Preface

This work had its origins in two projects. The first, undertaken in 1970 with Elmer Rees, was to construct certain low dimensional embeddings of real projective spaces. In order to do this we needed methods for calculating unstable homotopy groups of truncated projective spaces and associate spaces, as well as their images under various Freudenthal suspension homomorphisms. The second was to understand Mahowald's work on the metastable homotopy of spheres.

In 1971 and 1972 my work in surgery made me enlarge the scope of the project and consider an apparently unrelated problem - the stable homotopy of the Eilenberg-MacLane spaces $K(Q/Z,n)$. By means of appropriate fiberings these questions are seen to be merely different faces of the same coin.

Hence, this current work which provides a relatively effective framework for considering such questions. We generalize Mahowald's constructions to allow us to apply Adams' spectral sequence techniques to calculations, and we give detailed calculations for many examples; in particular, those needed for the work with Rees, and those needed in surgery with coefficients.

CONTENTS

Introduction

In recent years, stable homotopy theory has become a standard tool for the working topologist. If $\mathcal{Y}$ is a spectrum $\{Y_1, Y_2, \ldots, Y_n, \ldots\}$, one defines the stable homotopy groups of X with coefficients in $\mathcal{Y}$ as

$$H_i(X,Y) = \lim_{n\to\infty} \pi_{n+i}(X \wedge Y_n) ,$$

and these groups, according to G. W. Whitehead, define a generalized homology theory. (This means that they satisfy all the Eilenberg-Steenrod axioms except the dimension axiom.) In particular, if $\mathcal{Y}$ is the sphere spectrum $\mathcal{S} = \{S^1, S^2, \ldots, S^n, \ldots\}$, then

$$H_i(X, \mathcal{S}) = \pi_i{}^S(X)$$

defines the stable homotopy groups of the space X .

Since these groups form a homology theory, it is not surprising that homological techniques can be applied in calculations. Indeed, a major tool is the Adams spectral sequence, which has E^2-term

$$\mathrm{Ext}_{\mathcal{A}(p)}(H^*(X \wedge Y, Z_p), Z_p) ,$$

and converges to $H_*(X, \mathcal{Y} \otimes Z_{p^\infty})$. Here $\mathcal{A}(p)$ is the mod (p) Steenrod algebra, and $Z_{p^\infty} = \lim_{\to} Z_{p^i}$ (see e.g. [2]). Sometimes there are more efficient methods for calculating these groups, but in a range it is quite efficient, and has led to the calculation of the stable homotopy of spheres and some associated spaces through approximately the first 60 stems. This is a considerable achievement, since it was not that long ago that there was real uncertainty as to the order of even the second stem.

These techniques have also yielded spectacular results, such as Adams' first proof of Hopf-invariant 1 and his later solution of the vector field problem for spheres. Moreover, the recent work of Mahowald, Quillen and others on the structure of the stable J-homomorphism promises further profound knowledge in stable homotopy theory.

Summarizing, we can regard the stable groups as reasonably well understood.

For many problems, though, it is unstable groups which are actually needed. For example, the vector field problem for spheres (or its homotopy version) was really a question about how far back a certain element (the Whitehead product $[I,I]$) in $\pi_{4n-3}(S^{2n-1})$ desuspends. Its solution was obtained only after it was converted into a problem about the stable homotopy groups of truncated projective spaces, and the fact that this unstable question actually admitted such a reduction is, I suppose, the starting point of this monograph.

In a range of dimensions, there is an exact sequence, discovered in a special case by I. M. James and written down in generality by H. Toda, called the EHP sequence

$$\to \pi_i(S^n) \overset{E}{\to} \pi_{i+j}(S^{n+j}) \overset{H}{\to} \pi_{i-1}(S^{n-1} \wedge RP_n^{n+j-1}) \overset{P}{\to} \pi_{i-1}(S^n) \to ,$$

which relates the unstable homotopy groups of S^n, S^{n+j} to the (stable) groups of a truncated real projective space $(RP^m_n = RP^m/RP^{n-1} = *)$.

The existence of an element $\beta \in \pi_{4n-3-i}(S^{2n-i-1})$, which suspends to the Whitehead product $[I,I]$ can be interpreted as equivalent to the existence of an element $\alpha \in \pi_{4n-3-i}(S^{2n-i-2} \wedge P^{2n-1}_{2n-i-1})$, with $P(\alpha) = \beta$ in the EHP sequence, and, on pinching $S^{2n-i-2} \wedge P^{2n-2}_{2n-i-1}$ to a point, satisfies $p(\alpha) = e \in \pi_{4n-3-i}(S^{4n-3-i})$ is a generator, provided i is sufficiently small that the EHP sequence is valid. This can be seen at once on considering the commutative diagram of EHP sequences

$$\begin{array}{ccccccc}
\xrightarrow{P} \pi_{4n-3-i}(S^{2n-i-1}) & \to & \pi_{4n-2}(S^{2n}) & \xrightarrow{H} & \pi_{4n-3-i}(S^{2n-i-1} \wedge P^{2n-1}_{2n-i-1}) & \to \\
\downarrow E & & \downarrow = & & \downarrow E' & \\
\xrightarrow{P} \pi_{4n-2-i}(S^{2n-i}) & \xrightarrow{E} & \pi_{4n-2}(S^{2n}) & \xrightarrow{H} & \pi_{4n-2-i}(S^{2n-i} \wedge P^{2n-1}_{2n-i}) & \to \\
\downarrow E & & \downarrow = & & \downarrow E' & \\
\vdots & & \vdots & & \vdots & \\
\xrightarrow{P} \pi_{4n-3}(S^{2n-1}) & \xrightarrow{E} & \pi_{4n-2}(S^{2n}) & \to & \pi_{4n-2}(S^{2n-1} \wedge P^{2n-1}_{2n-2}) & \to
\end{array}$$

since $[I,I]$ generates the kernel of the bottom suspension map.

There are other problems of a similar nature involving, for example, the number of times a Thom space is really a suspension, which have implications for the geometric dimensions of vector bundles and the immersion dimensions for manifolds ([42]).

Our main object here is to develop machinery which leads to systematic methods for attacking such problems in a range of dimensions. Specifically, we study the problem of passing from the stable to the metastable homotopy groups of a space. In particular, we develop the following generalization of the EHP sequence.

<u>Theorem</u> 1.11. <u>There is a space</u> $\Gamma_L(X) = S^{L-1} \ltimes_T (X \wedge X)$, <u>which is</u> $(2n-1)$-<u>connected whenever</u> X <u>is</u> $(n-1)$-<u>connected and an exact sequence valid in the metastable range</u> $(i < 3n-2)$

$$* \qquad \cdots \xrightarrow{E} \pi_{i+L}(\Sigma^L X) \xrightarrow{H} \pi_i(\Gamma_L(X)) \xrightarrow{\partial} \pi_{i-1}(X) \xrightarrow{E} \pi_{i+L-1}(\Sigma^L X) \to \cdots .$$

So for L sufficiently large, $*$ determines $\pi_*(X)$ in terms of the <u>stable</u> homotopy of X and $\Gamma_L(X)$ in this range. If X is a sphere S^n , then $S^{L-1} \ltimes_T S^n \wedge S^n = \Sigma^n RP_n^{n+i-1}$ ([22]); however, for more complicated X, $\Gamma_L(X)$ becomes considerably more complex. In §§2 and 3, we give $H^*(\Gamma_L(X))$ as a module over the Steenrod algebra $\mathcal{A}(2)$ or $\mathcal{A}(p)$, provided we know the Steenrod algebra structure of $H^*(X)$. Thus, in principle, we can apply the Adams spectral sequence to determine $\pi_*(\Gamma_L(X))$ through a suitable range.

The idea behind the proof of 1.11 is contained in two facts: that $\pi_i(\Omega^n Y) \cong \pi_{n+i}(Y)$, where $\Omega^n Y$ is the nth loop space of Y, and that the natural map (§1)

$$j : X \hookrightarrow \Omega^n \Sigma^n X$$

gives, on passing to homotopy, the map $j_* : \pi_i(X) \to \pi_{n+i}(\Sigma^n X)$, which is the suspension map E in 1.11*. Converting j into a fibration and identifying the fiber with $\Omega(\Gamma_n(X))$ gives 1.11.

In order to make this identification, we need some basic facts about the structure of the space $\Omega^n \Sigma^n X$. When $n = 1$, I. M. James showed that $\Omega\Sigma X \simeq J_1(X)$, where $J_1(X)$ is the "reduced join"

$$(X \cup_{F_2} X^2) \cup_{F^3} X^3 \cup \ldots .$$

Here F_2 is defined on $* \times X \cup X \times *$ as the folding map, F_3 is defined on $* \times X^2 \cup X \times * \times X \cup X^2 \times *$ as the folding map, and so on. There is an associative product in $J_1(X)$ defined by juxtaposition, and $*$ is then the identity. In fact, $J_1(X)$ can be described as the free associative H-space generated by X with $*$ as unit. This result was generalized in [18] to give similar constructions for $\Omega^n \Sigma^n X$. We review (and explain) this construction in §0.

Specifically, we start almost from first principles and develop the geometric ideas which lead to an understanding of the basic structure of $\Omega^n\Sigma^n X$. These lead to constructions $J_1(X)$, $J_2(X) \ldots J_n(X)$, which are, in a sense, minimal models containing all the basic structures just developed. It is then a theorem that, for reasonable X, $J_n(X) \simeq \Omega^n\Sigma^n X$. We do not prove this last result (the proof can be found in [18]), but we do explain the considerations which lead to the constructions, and these should make the details in [18] almost unnecessary.

In §4, we consider the problem of looking at $\Omega^n X$ when X is no longer an n-fold suspension. The result is quite intriguing. There is an evaluation map $e : S^n \wedge \Omega^n X \to X$, and looping e n-times gives

$$\tau : \Omega^n\Sigma^n(\Omega^n X) \to \Omega^n X .$$

The fiber of τ is shown to be $\Gamma_n(X)$ in a range, and we obtain

<u>Theorem</u> 4.4. <u>Suppose</u> X <u>is an</u> (n+m-1)-<u>connected</u> CW-<u>complex</u>. <u>Then in dimensions less than</u> $3m-1$, $H^*(\Omega^n X, Z_p)$ <u>depends only on</u> $H^*(X,Z_p)$ <u>for</u> p <u>odd</u>, <u>and on</u> $H^*(X,Z_2)$ <u>as a module over</u> $\mathscr{A}(2)$ <u>for</u> $p = 2$.

To finish Part 1, we apply the results of §4 to the desuspension problem. The result is

<u>Theorem</u> 5.1. <u>Let</u> X <u>be</u> (n-1)-<u>connected and have dimension less than</u> $3n-2$. <u>Then</u>

1) <u>if</u> Y <u>is the</u> (2n-L-1)-<u>skeleton of</u> X, <u>there is a unique space</u> Z, <u>so</u> $\Sigma^L Z = Y$;
2) X <u>is itself an</u> L-<u>fold suspension if and only if a certain</u> (universally constructed) <u>map</u>

$$\phi : X/Y \to \Sigma^{L+1} S^{L-1} \ltimes_T Z \wedge Z$$

is homotopically trivial (as usual, ϕ is a stable map).

The result would be more satisfactory if we knew more about ϕ or even the cofiber of ϕ. In low dimensions, things can be explicitly worked out using unstable higher cohomology operations (see e.g. [36]), but at present the author has no general results.

In this connection, we would like to point out the worked example at the end of §4, $\Omega^{11}(CP^{12}{}_8)$, where we show that the mod 2 Steenrod algebra action in $H^*(\Omega^n X)$ is not determined by its action in X. It is possible to interpret the work of Adém-Gitler on non-immersion theorems ([3]) in terms of examples of this kind, and such analysis could lead to a sharper understanding of ϕ.

In Part 2, we develop means for calculating the maps H and P in 1.11*.

Mahowald showed in [13] how to use Adams spectral sequence techniques to study the ordinary EHP sequence. He constructed a map

$$E^2(H) : \mathrm{Ext}^{s,t}_{\mathcal{A}(2)}(Z_2, Z_2) \to \mathrm{Ext}^{s-1,\, t-n-1}_{\mathcal{A}(2)}(H^*(RP_n), Z_2),$$

which commutes with differentials and, at E^∞, gives a map associated with H. (Note here that $E^2(H)$ changes the s degrees. It is precisely this change which makes Mahowald's map non-trivial.)

The obvious generalization of $E_2(H)$ to the map in 1.11* fails, however, and $E^2(H)$ does not exist for any space more complicated than a sphere (at the prime 2) !

Our main object in Part 2 is to provide a satisfactory generalization. We first review Adams' method for constructing his spectral sequence, and generalize it slightly so as to define an unstable spectral sequence which approximates the actual homotopy of a space X. Convergence seems difficult in general, but the sequence does converge in the metastable range. There is a natural suspension map $\hat{E}$ from this spectral sequence to the stable Adams spectral sequence, which is an isomorphism in the stable range and, at E^∞, is associated to E. At the E^2 level, $\hat{E}$ is algebraically determined through the metastable range. This situation is quite nice except that the E^2-term of our sequence for X is very hard to determine above the stable range, so we turn to the maps H and P.

In 6.11, we indicate how to define a spectral sequence for a pair (Y,A), where $A \subset H^*(Y,Z_2)$ is any submodule closed under the action of $\mathscr{A}(2)$. The resulting modified Adams spectral sequence has E^2-term calculable in terms of $\mathrm{Ext}_{\mathscr{A}(2)}(A,Z_2)$, $\mathrm{Ext}_{\mathscr{A}(2)}(H^*(Y)/A, Z_2)$, and a differential

$$\partial : \mathrm{Ext}^{i,*}(H^*(Y)/A, Z_2) \to \mathrm{Ext}^{i+2,*}(A,Z_2) .$$

We denote it $E^2_{**}(Y,A)$.

In $H^*(\Gamma_L(X), Z_2)$, there is a natural $\mathscr{A}(2)$-submodule A, and in §8 we construct a map

$$\partial^2 : \mathrm{Ext}^{s,t}_{\mathscr{A}(2)}(H^*(X), Z_2) \to E^2_{s-1,t-2}(\Gamma_L(X), A) ,$$

which provides the desired generalization of Mahowald's map $\hat{E}^2$ (A is 0 if and only if X is a sphere at 2, in which case $E^2_{**}(\Gamma_L(X), 0) = \mathrm{Ext}^{*,*-n-1}_{\mathscr{A}(2)}(H^*(\Gamma_L(X), Z_2)$. In fact, we are able to prove

<u>Theorem</u> 8.5. <u>Suppose</u> $L > 3m$, <u>with</u> X $(m-1)$-<u>connected</u>. <u>Then there is a map</u>

$$J^* : E^*_{i,j}(\Gamma_L(X), A) \to E^*_{i+1,j+1}(X) ,$$

<u>and, in the metastable range, the sequence</u>

$$** \qquad \ldots \to \mathrm{Ext}^{s,t}_{\mathscr{A}(2)}(H^*(X), Z_2) \xrightarrow{\partial^2} E^2_{s,t-1}(H^*(\Gamma_L(X)), A)$$

$$\xrightarrow{J^2} E^2_{s+1,t}(X) \xrightarrow{\hat{E}^2} \mathrm{Ext}^{s+1,t}_{\mathscr{A}(2)}(H^*(X), Z_2) \to \ldots$$

<u>is long exact</u>.

This result makes effective calculations feasible in some cases. To expedite them, we conclude Part 2 with a discussion (§9) of methods for calculating $\mathrm{Ext}^{i,j}_{\mathscr{A}(2)}(H^*(\Gamma_L(X)/A, Z_2)$. This is highly non-trivial in general, since $H^*(\Gamma_L(X)/A)$ is a very complex $\mathscr{A}(2)$-module; it replaces each Z_2-cohomology class of X by the cohomology of a truncated projective space. By an appropriate filtration, we obtain a spectral sequence converging to $\mathrm{Ext}_{\mathscr{A}(2)}(H^*(\Gamma_L(X)/A), Z_2)$, whose E^1-term contains a copy of $\mathrm{Ext}^{**}_{\mathscr{A}(2)}(H^*(RP_n), Z_2)$ for each n-dimensional cohomology class in $H^*(X,Z_2)$. A second spectral sequence is also developed, which makes calculations feasible in case $\mathrm{Ext}_{\mathscr{A}(2)}(H^*(X), Z_2)$ is sufficiently well-known.

In Part 3, we apply the results of Parts 1 and 2, and give some examples to show that the theorems there cannot be improved too much.

In §§10 and 11, we calculate some of the stable homotopy of the Eilenberg-MacLane spaces $K(Z,n)$, $K(Z_2,n)$, and $K(Q/Z, n)$. The results for $K(Q/Z, n)$ are calculated only so far as we need them in applications ([41])

$$\pi^s_{2n}(K(Q/Z, n)) = 0 ,$$

$$\pi^s_{2n+1}(K(Q/Z, n)) = \begin{cases} Q/2Z , & n\text{-odd} \\ 0 , & n\text{-even} \end{cases} .$$

However, in §11, we give the first 10 stable groups for $K(Z,8k+1)$ as an example (Theorem 11.18), and do most of the necessary work to obtain these groups for other values of n as well. In particular, 11.18 corrects some errors in [14].

§12 applies the metastable sequence 8.5** to the case $X = S^n \cup_2 e^{n+1}$, and, as an example, we calculate the homotopy groups of $S^7 \cup_2 e^8$ through the entire metastable range.

Finally, in §13, we give explicit calculations for some truncated projective spaces. In particular, our final calculation is of the unstable resolution for P^6_2 to slightly beyond the metastable range, where we see that wild filtration changes make any reasonable extension of 8.5** impossible. The remaining calculations in §13 provide the homotopy theoretic results needed in [42].

These results were originally obtained in 1969 and 1970. Since then, there has been further work by several authors on some of the questions considered here. B. Drachman has studied the desuspension problem from another point of view, and has obtained geometric criteria for deciding when a space is a suspension. Unfortunately, it seems difficult to iterate his techniques.

Also, an area which has received only partial attention but clearly merits more is the extension of the current results to generalized homology theories, such as unstable MU-theory, which seems ready for serious development in view of W. Steven Wilson's thesis (M.I.T., 1972).

In this connection, it would be interesting to explain Donald Davis' thesis (Stanford, 1971) on the geometric dimensions of bundles over RP^n in terms of obstructions to desuspension of the Thom complexes, since that would probably give insight into the nature of the map ϕ (5.1) in b_O or b_U theory.

P A R T   I

§0. Iterated loop spaces

We begin by describing the category $\mathscr{C}_n$ of n-fold loop spaces. We can look at $\Omega^n X$ as the space of base point-preserving maps $S^1 \to \Omega^{n-1}X$ or the base point-preserving maps $S^2 \to \Omega^{n-2}X \ldots$, or $S^n \to X$. With respect to the various ways of looking at $\Omega^n X$, there are evaluation maps

$$\mathrm{adj}^n_k(1) : S^k \wedge \Omega^n(X) \to \Omega^{n-k}(X) ,$$

which fit together to give

0.1
$$S^n \wedge \Omega^n(X) \xrightarrow{\Sigma^{n-1}\mathrm{adj}^n_1(1)} S^{n-1} \wedge \Omega^{n-1}(X) \xrightarrow{\Sigma^{n-2}\mathrm{adj}^{n-1}_1(1)} S^{n-2} \wedge \Omega^{n-2}(X) \to \ldots \to X$$

and the composite

$$\Sigma^{n-s}\mathrm{adj}^{n-s+1}_1(1) \ldots \Sigma^{n-1}\mathrm{adj}^n_1(1) = \Sigma^{n-1}\mathrm{adj}^n_s(1) .$$

The Moore loop space $\Omega_{(M)}(X)$ is the set of maps $f : [0,k_f] \to X$ for variable $k \geq 0$, which satisfy $f(0) = f(k) = *$. It has an associative product with unit $f * g : [0, k_f + k_g] \to X$, defined by setting

$$f * g(t) = \begin{cases} f(t) , & t \leq k_f \\ f(t-k_f) , & t \geq k_f . \end{cases}$$

The unit of $\Omega_M(X)$ is $* : [0] \to *$. Also, there is a natural homotopy equivalence between $\Omega(X)$ and $\Omega_M(X)$ (as, for example, in J. Adams and P. Hilton, "On the chain algebra of a loop space," Comment. Math. Helv. 30 (1956), 305-330), so in the remainder of this paper, we identify them and leave it to the reader to make the necessary modifications to go from $\Omega^n(X)$ to $\Omega^n{}_M(X)$ or vice versa.

If Z and W are n-fold loop spaces, then a map $f : Z \to W$ is "admissible" if and only if $f = \Omega^n(g)$ for some $g : \Omega^{-n}(Z) \to \Omega^{-n}(W)$.

Lemma 0.2. _Let_ $f : X \to \Omega^n Y$ _be a map. Then there is an admissible map_

$$g(f) : \Omega^n \Sigma^n X \to \Omega^n Y$$

and a natural inclusion

$$j : X \to \Omega^n \Sigma^n X ,$$

_so_ $g(f) \cdot j = f$.

Proof. From the composite $\Sigma^n X \xrightarrow{\Sigma^n f} \Sigma^n \Omega^n Y \xrightarrow{adj(1)} Y$, we can loop down n-times, obtaining the map $g(f)$. Now $j : X \to \Omega^n \Sigma^n X$ is defined by $j(x)(\vec{t}) = (\vec{t}, x) \in S^n \wedge X$, and 0.2 follows.

Remark 0.3. $\pi_n(Y) \cong \pi_{n-k}(\Omega^k Y)$, $n \geq k$. Also, if $Y = \Sigma^k Z$, then the map

$$j : Z \to \Omega^k \Sigma^k Z$$

gives $j_* : \pi_s(Z) \to \pi_{k+s}(\Sigma^k Z)$, and this is the Freudenthal suspension homomorphism. If Z is an (n-1)-connected CW-complex, then the Freudenthal suspension theorem implies $j : Z \to \Omega^k \Sigma^k Z$ is a homotopy equivalence in dimensions less than $2n-1$, since Milnor has shown that $\Omega^k \Sigma^k Z$ also has the homotopy type of a CW-complex ([25]).

Remark 0.4. The universal example for the situation in 0.2 is $g(id) : \Omega^n \Sigma^n (\Omega^n Y) \to \Omega^n Y$. Indeed, it has recently been shown by J. P. May that the existence of an H-map $\Omega^n \Sigma^n Z \xrightarrow[\pi]{} Z$, with $\pi \cdot j = id$, is essentially equivalent to the associative H-space Z being an n-fold loop space.

These observations signal the basic role of the spaces $\Omega^n\Sigma^n Y$ in $\mathcal{C}_n$. We study the map j in more detail in §1 and $\tau = g(1)$ in §4. Now our object is an explicit description of the homotopy type of the space $\Omega^n\Sigma^n X$.

A model for $\Omega\Sigma X$ was constructed by I. M. James (in "Reduced product spaces," Ann. of Math. 62 (1955), 170-197). It is easily described. Set $J_1(X) = \cup_{n=1}^{\infty} X^n/R$, where R is the relation

0.5 $$(x_1 \ldots x_i, *, x_{i+2} \ldots x_n) \sim (x_1 \ldots x_i, x_{i+2} \ldots x_n) .$$

It has an associative product (juxtaposition), a unit $*$, an obvious topology, and James proved that $J_1(X) \simeq \Omega\Sigma X$ for X a CW-complex. The equivalence of $J_1(X)$ with $\Omega\Sigma X$ is obtained by mapping $J_1(X) \to \Omega\Sigma X$ as the (unique) multiplicative extension of $j : X \to \Omega_M(\Sigma X)$.

Models for $\Omega^n\Sigma^n X$, $n > 1$, were constructed in [18]. Several attempts to rework the construction occurred thereafter, culminating with the construction given by J. P. May (in The Geometry of Iterated Loop Spaces, Lecture Notes in Mathematics 271, Springer-Verlag, 1972). In 1.14, we describe the basic germ of his models, but in the remainder of this section, we largely follow [18].

Let us begin by looking at $J_1(\Sigma Y) \simeq \Omega\Sigma^2 Y$. Its component building blocks, the $(\Sigma Y)^n$ can be written after shuffling the variables in the form

$$I^n \times Y^n ,$$

where we make the identifications

$$(t_1 \ldots \varepsilon_i \ldots t_n, y_1 \ldots y_n) \sim (t_1 \ldots t_n, y_1 \ldots, y_i = *, \ldots y_n)$$

$$\sim (t_1 \ldots \hat{t}_i \ldots t_n, y_1 \ldots \hat{y}_i, \ldots y_n), \qquad n > 1,$$

and $(\varepsilon_1, y_1) \sim (t_1, *) \sim *$ for $n = 1$. Here $\varepsilon_i = 0$ or 1.

Let $P(n)$ be the set of variable length paths starting at $(0,0,\ldots,0)$ in I^n and ending at $(1,\ldots,1)$. Crossing with Y^n, we map $P(n) \times Y^n$ into paths on $I^n \times Y^n$, starting at $(0,0,\ldots,0) \times Y^n$ and ending at $(1,\ldots,1) \times Y^n$, by defining $(f, y_1 \ldots y_n)t = (f(t), y_1 \ldots y_n)$.

In view of our identification 0.6, we see that $(0,\ldots,0) \times Y^n \sim (1,\ldots,1) \times Y^n \sim *$ in $J_1(\Sigma Y)$. Thus, in $J_1(\Sigma Y)$, *each point of* $P(n) \times Y^n$ *gives rise to a loop*; i.e., there is a well-defined and continuous map

$$\phi_n : P(n) \times Y^n \to \Omega J_1(\Sigma Y) \simeq \Omega^2 \Sigma^2 Y .$$

As a first approximation of $\Omega^2\Sigma^2 Y$, we could take the free associative H-space generated by the disjoint union of the $P(n) \times Y^n$, and extend the ϕ_n to a multiplicative map in the evident way. However, to do this would be to overlook at least one vital bit of additional structure in the ϕ_n.

Definition 0.7. *There is a pairing*

$$u_{i,j} : P(i) \times P(j) \to P(i+j),$$

defined by

$$u_{i,j}(f,g)t = \begin{cases} (f(t), 0,\ldots 0), & t \le |f| \\ (1,\ldots,1, g(t-|f|)), & t \ge |f|, \end{cases}$$

where $|f|$ *is the length of the path* f.

Clearly, the $u_{i,j}$ are associative in the sense that

$$u_{i+j,k}(u_{i,j} \times 1) = u_{i,j+k}(1 \times u_{j,k}) .$$

Lemma 0.8. The map

$$(P(n) \times Y^n) \times (P(m) \times Y^m) \xrightarrow{\phi_n \times \phi_m} \Omega(J_1(\Sigma Y)) \times \Omega(J_1(\Sigma Y)) \xrightarrow{u} \Omega(J_1(\Sigma Y))$$

factors as the composite

$$(P(n) \times Y^n) \times (P(m) \times Y^m) \to P(n) \times P(m) \times (Y^n \times Y^m) \xrightarrow{u_{m,n} \times 1} P(n+m) \times Y^{n+m} \xrightarrow{\phi_{n+m}} \Omega(J_1(\Sigma Y)) .$$

(The proof is obvious.)

Thus a better model for $\Omega J_1(\Sigma Y)$ would be obtained from the union of the $P(n) \times Y^n$ by making a further identification

0.9 $$(p,y)\cdot(p',y') \sim (u_{n,m}(p,p'), (y,y')) .$$

The resulting model, although better, is still too big. Recall that, in 0.6, if $y_i = *$ in Y^n, we collapse $I^n \times Y^n$ on $I^{n-1} \times Y^{n-1}$ by forgetting the ith coordinates.

Lemma 0.10. The map $\lambda_i : I^n \to I^{n-1}$, forgetting the ith coordinate, induces a map

$$P(\lambda_i) : P(n) \to P(n-1) ,$$

and if

$$\psi_n : P(n) \times Y^n \to \Omega((\Sigma Y)^n)$$

is the obvious map, $j_n : (\Sigma Y)^n \to J_1(\Sigma Y)$ *the quotient map, then*

$$\phi_n = j_n \psi_n ,$$

and the diagram

$$\begin{array}{ccc} P(n) \times Y^n & \xrightarrow{\psi_n} & \Omega[(\Sigma Y)^n] \\ {\scriptstyle P(\lambda_i)\times\lambda_i}\downarrow & & \downarrow{\scriptstyle \Omega\lambda_i} \\ P(n-1) \times Y^{n-1} & \xrightarrow{\psi_{n-1}} & \Omega[(\Sigma Y)^{n-1}] \end{array}$$

commutes.

Thus we can add another relation to our construction:

0.11 $$(p,y) \sim (P(\lambda_i)(p), \lambda_i(y)) \text{ if}$$

$$y = (y_1 \ldots y_{i-1}, *, y_{i+1} \ldots y_n) .$$

Finally, there is one more type of relation which must be taken into account. It is well-known that a second loop space has a *homotopy commutative multiplication*. We add homotopy commutativity to our model as follows.

Definition 0.12. *The symmetric group* $\mathcal{S}_n$ *acts on* $P(n)$ *by* $\alpha(p)(\vec{t}) = p(\alpha\vec{t})$, *where* $\alpha(t_1, \ldots, t_n) = (t_{\alpha^{-1}(1)}, \ldots, t_{\alpha^{-1}(n)})$.

Lemma 0.13. $P(n)$ *is equivariantly contractible with respect to the* $\mathcal{S}_n$*-action* (i.e., the homotopy of contraction can be chosen so that $H_t(\alpha p) = \alpha H_t(p)$, all $p \in P(n)$) .

Proof. We start by defining a contraction of I^n by $\ell_t(t_1 \dots t_n) = (tt_1, \dots, tt_n)$, and corresponding to ℓ_t, the contraction H_t is given by

$$H_t(f)(\tau) = \begin{cases} \ell_t f(\tau/t), & \tau \le t|f| \\ (t+\tau-t|f|, \dots, t+\tau-t|f|) & \text{otherwise.} \end{cases}$$

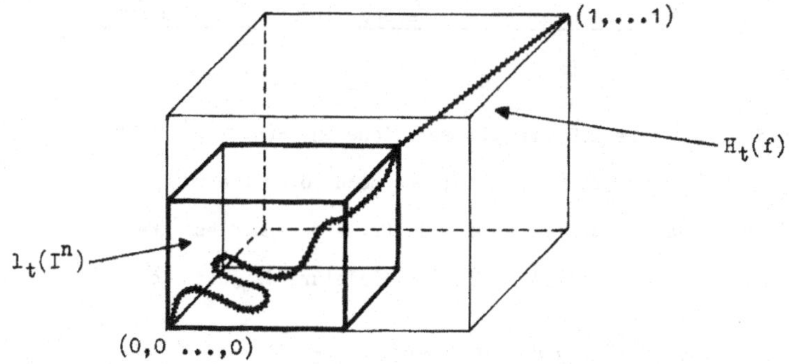

Similarly, we have

Lemma 0.14. The following diagram commutes

$$\begin{array}{ccc}
(P(n) \times Y^n) \times (P(m) \times Y^m) & \xrightarrow{\phi_n \times \phi_m} & \Omega(J_1 \Sigma Y) \times \Omega(J_1 \Sigma Y) \\
\downarrow{\scriptstyle (u_{n,m} \times 1)(\text{shuff})} & & \downarrow{\scriptstyle T} \\
P(n+m) \times Y^{n+m} & & \Omega(J_1 \Sigma Y) \times \Omega(J_1 \Sigma Y) \\
\downarrow{\scriptstyle s_{n,m} \times 1} & & \downarrow{\scriptstyle u} \\
P(n+m) \times Y^{n+1} & \xrightarrow{\phi_{n+m}} & \Omega J_1(\Sigma Y)
\end{array}$$

where $s_{n,m} \in \mathscr{S}_{n+m}$ is the shuffle of the first n with the last m-coordinates, and T is the interchange.

In particular, this implies the additional identification

0.15 $$(s_{n,m}(u_{n,m}(p,p')), (y,y')) \sim (u_{m,n}(p',p), y', y) .$$

But $(u_{m,n}(p',p), y', y)$ is equivalent by 0.9 to the product $(p',y')(p,y)$, and since $P(n)$ is connected, $(s_{n,m}(u_{n,m}(p,p')),y,y')$ is homotopic to $(u_{n,m}(p,p'), (y,y')) \sim (p,y)\cdot(p',y')$.

Theorem 0.16. *Let* $K_2(Y) = \cup_n P(n) \times Y^n$ *modulo the relations* 0.9, 0.11, *and* 0.15. *Then* $K_2(Y)$ *is an associative H-space with unit, and the natural map* $K_2(Y) \to \Omega J_1(\Sigma Y)$ *is a homotopy equivalence if* Y *is a connected CW-complex.*

Proof. Consider the type 0.9 and 0.15 relations on $P(n) \times Y^n$. They imply that the only type of $(P(n))$-relations occur over the points $\alpha(P(r) \times P(n-r))$, where α runs over all $(r,n-r)$ shuffles. We call these the $n-2$ "faces" of $P(n)$. This nomenclature is justified by

Lemma 0.17. *Let* $F = \cup_{\alpha,r=1}^{n-1} \alpha(P(r) \times P(n-r))$. *Then*

i) $H_\ell(P(n),F;\ Z) = \begin{cases} 0\ , & \ell \neq n-1 \\ Z\ , & \ell = n-1 \end{cases}$.

ii) *Let a generator* p_{n-1} *of* $H_{n-1}(P(n), F)$ *be given. Then the evaluation map*

$$e : (I,\partial) \times (P(n),F) \to (I^n, \partial I^n)$$

has degree ± 1 ; i.e.,

$$e_*(e_1 \otimes p_{n-1}) = \pm e_n ,$$

where e_1 *is the orientation class of* (I^1,∂).

(The proof is by induction.)

In J. F. Adams' paper, "On the cobar construction," Proc. Nat. Acad. Sci. U.S.A. 42 (1956), 409-412, it is shown that there is a spectral sequence defined for any space X, and converging to $H_*(\Omega X, Z_p)$. Its E^2-term has the form $\mathrm{Cotor}_{H_*(X,Z_p)}(Z_p, Z_p)$. A similar spectral sequence can be defined for $K_2(Y)$, and the natural map $K_2(Y) \to \Omega(J_1(\Sigma Y))$ induces a map of spectral sequences. From 0.17, it is then an easy calculation to check that, at the E^2-level, the spectral sequence map is actually an isomorphism. Then if Y is a finite complex, the comparison theorem shows the natural map induces isomorphisms in homology for all coefficients Z_p, and it is known that $\Omega^2\Sigma^2 Y$ has the homotopy type of a locally finite CW-complex. Hence by the Whitehead theorem (for connected H-spaces), the natural map is a homotopy equivalence. Now, since Y is the limit of its finite subcomplexes, the result follows for general Y.

Notice the role of the "complex" of faces $\alpha(P(r) \times P(k-r))$, $\alpha((\beta P(s) \times P(r-s)) \times P(n-r))$, etc., in the proof of 0.16. Through 0.17, they are the essential things in making the proof work. The complex for $P(2)$ is simply that of an interval; that for $P(3)$ has the form of a hexagon

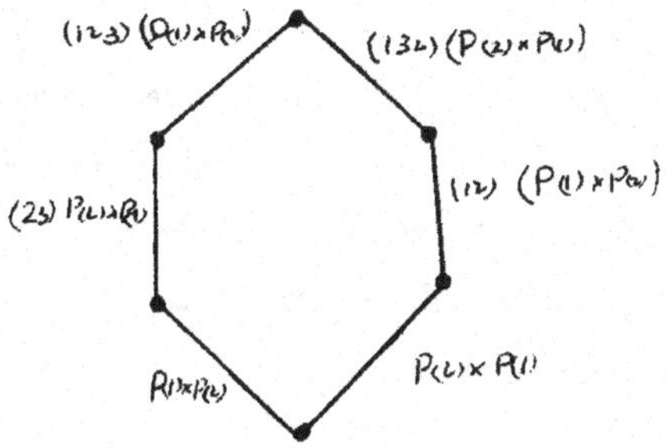

As these examples indicate, the complexes above are obtained from cellularly decomposing the boundary of a disc D^{n-1} , and we may simply replace $P(n)$ by D^{n-1} , and make the type 0.9 and 0.15 identifications over the corresponding faces to give a much smaller model for $\Omega^2\Sigma^2 X$.

The explicit construction follows. Recall first that the faces of the convex hull of a finite point set S in Euclidean space are convex hulls of certain subsets of S .

Definition 0.18. *Let* $C(n)$ *be the convex hull of the translates of* $(1,2,3,\ldots,n+1) \in R^{n+1}$ *under the action of* $\mathcal{S}_{n+1}$.

$C(n)$ is easily seen to be the hexagon for $n = 2$, the figure

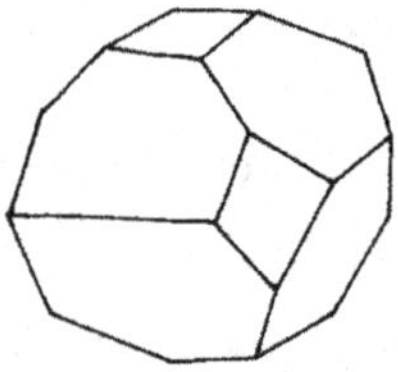

whose faces consist of eight hexagons and six squares $C(1) \times C(1)$ for $n = 3$, and in general $C(n)$ has faces of the form $\alpha(C(r) \times C(n-r-1))$, $r = 0, \ldots, n-1$, as α runs over all $(r+1, n-r)$ shuffles. Indeed, let $S' \subset S$ be the orbit of $(1,2,\ldots,n+1)$ under the action of $\mathcal{S}_{r+1} \times \mathcal{S}_{n-r}$. Then the convex hull of S' is naturally isomorphic to $C(r) \times C(n-r-1)$. Similarly, $\alpha C(r) \times C(n-r-1)$ is the convex hull of $\alpha(S')$. (See Lemma 4.2, p. 391 of [18] for details.) Let $I^r : C(r) \times C(n-r-1) \to C(n)$ be the identification above.

Note that $C(n)$ is invariant under the action of $\mathscr{S}_{n+1}$ which takes faces to faces. Also, guided by the need for degeneracies in 0.10, 0.11, we can define degeneracies

$$D_i : C(n) \to C(n-1)\ , \quad i = 1\ , \ \dots\ , \ n+1$$

(as in Lemma 4.5, p. 392 of [18]). Specifically, we have

Lemma 0.19 ([18]). *There are maps*

$$D_k : C(n) \to C(n-1)\ , \qquad 1 \le k \le n+1\ ,$$

so that

i) $D_1 I^0$ *is the projection on the second factor;*

ii) $D_j I^k = \begin{cases} I^{k-1}(D_j \times id)\ , & j \le k \\ I^k\,(id \times D_{j-k}) & \textit{otherwise;} \end{cases}$

iii) $d_j(\beta) D_{\beta^{-1}(j)} = D_j \beta\ , \quad \beta \in \ _{n-1}\ ;$

iv) $D_i D_j = D_{j-1} D_i$ *for* $j \ge i$.

(Here $d_j : \mathscr{S}_{n+1} \to \mathscr{S}_n$ is the correspondence which makes the diagram

$$\begin{array}{ccc} R^{n+1} & \xrightarrow{\ \beta\ } & R^{n+1} \\ {\scriptstyle \lambda_{\beta^{-1}(j)}} \big\downarrow & & \big\downarrow {\scriptstyle \lambda_j} \\ R^n & \xrightarrow{\ d_j(\beta)\ } & R^n \end{array}$$

commute.)

Indeed, (i)-(iii) specify D_j on faces, and the map D_j is then defined by extending linearly along rays to the respective centers. The reader is advised to check that (i)-(iv) are forced on us by 0.10.

As the remarks preceding 0.18 indicate, the $C(n-1)$ serve as replacements for the $P(n)$, and we have

<u>Corollary</u> 0.20. $$\Omega^2\Sigma^2 X \simeq \bigcup_{n=0}^{\infty} C(n) \times X^{n+1}$$

<u>modulo the relations</u>

i) $(c, x_1 \ldots x_i *, x_{i+2} \ldots x_{n+1}) \sim (D^{i+1}(c), x_1 \ldots \hat{x}_{i+1} \ldots x_{n+1})$;

ii) <u>if</u> $c \in \alpha I^i(C(i) \times C(n-i-1))$ <u>for</u> α <u>an</u> $(i+1, n-1)$ <u>shuffle, then</u>

$$(c, x_1 \ldots x_{n+1}) \sim (\alpha^{-1}(c), x_{\alpha^{-1}(1)}, \ldots, x_{\alpha^{-1}(n+1)}),$$

<u>if</u> X <u>is a connected</u> CW-<u>complex with</u> $*$ <u>a vertex. The multiplication in</u> $J_2(X) = \cup_0^\infty C(n) \times X^{n+1}/R$ <u>is given as</u>

$$\{c, x_1 \ldots x_{n+1}\} \cdot \{c', x_{n+1} \ldots x_{n+m+2}\} = \{I^n(c, c'), x_1 \ldots x_{n+m+2}\},$$

<u>and with this product,</u> $J_2(x)$ <u>is</u> H-<u>equivalent to</u> $\Omega^2\Sigma^2 X$.

This provides us with a <u>minimal</u> geometric model for $\Omega^2\Sigma^2 X$. Basically, the model makes it clear that the fundamental data which go into the statement that a space is a 2-fold loop space are an associative unitary multiplication, together with a series of higher homotopies of commutation.

We now indicate how to extend this construction to obtain models for the higher loop spaces $\Omega^n\Sigma^n X$, $n > 2$.

Consider, for example, the case $n = 3$. We have already approximated $\Omega^2\Sigma^2(\Sigma X)$ as $J_2(\Sigma X) = \cup C(n) \times (\Sigma X)^{n+1}/R = \cup C(n) \times I^{n+1} \times X^{n+1}/R' = \cup I^{n+1} \times C(n) \times X^{n+1}/R''$. Once again we can use the $P(n+1)$ to construct loops by taking elements of $P(n+1) \times C(n) \times X^{n+1}$ and defining

$$\phi(p,c,x_1\ldots x_{n+1})t = \{p(t),c,x_1\ldots x_{n+1}\} .$$

By 0.20(i), these paths become loops. Moreover, the obvious analogues of 0.8, 0.10, 0.14 continue to hold. Thus we obtain a model for $\Omega^3\Sigma^3X$ as follows.

Definition 0.21. *Set* $J_3(X) = \cup\,(C(n) \times C(n) \times X^{n+1})$ *modulo the relations*

i) $(c,c',x_1\ldots *_i,\ldots x_{n+1})$ $(D^i(c),D^i(c'),x_1\ldots \hat{x}_i,\ldots x_{n+1})$;

ii) *if* $c' \in \alpha I^k(C(k) \times C(n-k-1))$, *then*

$$(c,c',x_1\ldots x_{n+1}) \sim (\alpha^{-1}c,\alpha^{-1}c',\alpha(x_1\ldots x_{n+1})) ;$$

iii) *if* $c \in \alpha I^k(C(k) \times C(n-k-1))$, *then*

$$(c,c',x_1,\ldots,x_{n+1}) \sim (\alpha^{-1}c,D^\alpha(c'),\alpha(x_1\ldots x_{n+1})) .$$

Here $D^\alpha(c') = [(D_{k+2})^{n-k} \times (D_1)^{k+1}\alpha^{-1}](c')$ *is the obvious degeneracy used for paths on* $(\Sigma X)^{n+1}$, *which are at* $*$ *on the first* $k+1$ *coordinates half the time, and* $*$ *on the last* $n-k$ *coordinates the remainder of the time.*

To complete the definition, we remark that, after using 0.2(i)-(iii), *if the point is equivalent to one of the form* $((c_1,c_2),(c_1',c_2')x_1\ldots x_{n+1}x_{n+2}\ldots x_{n+m+2})$, *that is, to a product, then we again apply* (i)-(iii) *separately to* $(c_1,c_1'x_1\ldots x_{n+1})$, $(c_2c_2'x_{n+2}\ldots x_{n+m+2})$ *and take the product of the results.*

It is now possible to prove an analogue of 0.16 for this model, and we have

Theorem 0.22. *If* X *is a connected* CW-*complex, then* $J_3(X)$ *has the homotopy type of* $\Omega^3\Sigma^3 X$ *as an associative* H-*space*.

It is now clear how to generalize the construction. We obtain

Definition 0.23. *Set* $J_m(X) = \cup_{0=n}^{\infty} (C(n))^{m-1} \times X^{n+1}$ *modulo the relations*

i) $(c_1 \ldots c_{m-1}, x, \ldots *_i, \ldots x_{n+1}) \sim (D^i(c_1), \ldots, D^i(c_{m-1}), x, \ldots \hat{x}_i \ldots x_{n+1})$;

ii) *if* $c_j \in \alpha I^k(C(k) \times C(n-k-1))$, *then*

$$(c_1 \ldots c_{m-1}, x_1 \ldots x_{n+1}) \sim (\alpha^{-1}c_1, \ldots, \alpha^{-1}c_{j-1}, D^{\alpha}(c_j) \ldots D^{\alpha}(c_{m-1}), \alpha(x, \ldots, x_{n+1})) ;$$

iii) *the same convention on points equivalent to products as given in* 0.21.

Once again we can prove

Theorem 0.24 (Theorem 5.2 of [18], p. 395). *Let* X *be a connected* CW-*complex. Then there is an* H-*map*

$$j_k : J_k(X) \to \Omega^k\Sigma^k(X) ,$$

which is a homotopy equivalence.

Remark 0.25. There is a filtration on the points of $J_k(X)$, given by saying y has filtration m if it is in the equivalence class of a point in $(C(m-1)^{k-1} \times X^m)$ under the relations of 0.23. In particular, if $*$ is a vertex of X , and X is $(\ell-1)$-connected, then the set of points having filtration $\leq s$, $J_k(X)^{(s)}$, is a subcomplex of $J_k(X)$ and contains all the cells of $J_k(X)$ of dimension $< (s+1)\ell$ (provided X has no cells of dimension $< \ell$ except $*$, which we can assume). Thus if we wish to consider problems dealing with dimensions $\leq (s+1)\ell-2$, we can replace $J_k(X)$ by $J_k(X)^{(s)}$.

The reader is advised to work out the explicit structure of $J_k(X)^{(2)}$ and verify the description given in the proof of 1.11.

§1. The Inclusion $X \to \Omega^n\Sigma^n X$

Recall from §0 that the inclusion $j: X \hookrightarrow \Omega^n\Sigma^n X$ is defined by sending $x \in X$ to the map $f : (S^n,*) \to \Sigma^n X$, defined as the composition $S^n \to (S^n,x) \hookrightarrow S^n \wedge X = \Sigma^n X$. It is clearly natural and continuous, and in homotopy induces the map j_* : $\pi_j(X) \to \pi_{n+j}(\Sigma^n X)$, which is just the n-fold iterate of the Freudenthal suspension homomorphism. In particular, if X is an m-1-connected locally finite CW complex, then j is 2m-2-connected.

We convert j into a fibering in the usual way. Thus we first replace $\Omega^n\Sigma^n X$ by the mapping cylinder $M(j)$, and j by the inclusion $X \hookrightarrow M(j)$. Then F_n, the fiber of j, is defined to be the space of paths of unit length $E_X^{M(j)}{}_*$ starting in X and ending at $*$, the base point in $\Omega^n\Sigma^n(X)$. By the result of Milnor ([25]), if X is a CW complex, then F_n is the homotopy type of one also.

In [18], it was shown that $H_*(\Omega^n\Sigma^n(X), Z_p)$ is an explicit functor of $H_*(X,Z_p)$ alone, and the inclusion j_* is injective in homology. Thus, using the Serre spectral sequence of the fibering, it is easily argued that $H_*(F_n)$ depends only on $H_*(X)$ and n in dimensions less than $3m-1$ if X is m-1-connected.

Here is an alternate description of F_n. Let G, H be associative H-spaces with units, and $f : G \to H$ an inclusion which is also a homomorphism. f induces an inclusion of classifying spaces ([19], [32])

$$B_f : B_G \to B_H ,$$

and we have

Lemma 1.1 *Let* $E_H \xrightarrow{\rho} B_H$ *be the universal quasi-fibering* ([7], [19], [32]). *Then the fiber of* B_f *is* E_H *restricted to* B_G ; i.e., $\rho^{-1}(B_G)$.

The proof is direct.

In particular, if $G = \Omega(X)$ and $H = \Omega(M(\mathcal{J}))$, this provides an explicit and fairly manageable description of F_n . If $X = \Sigma Y$, the classifying space constructions given above can be considerably improved. Indeed, using the constructions introduced in [18], the inclusion $\Omega\Sigma Y \hookrightarrow \Omega^{n+1}\Sigma^{n+1}Y$ is H-equivalent to the inclusion

$$\lambda : J_1(Y) \hookrightarrow J_{n+1}(Y) .$$

In §2 of [18], an alternate classifying space construction is given: Let X be a (free[1]) associative H-space with unit $*$ and homomorphism $h : X \to R^+$ so $h^{-1}(0) = *$; then E_X is defined as $X \times R^+ \times X$ mod the relations

$$(x,t,yz) \sim (xy,t-h(y),z) ,$$
$$(x,0,y) \sim (x,t,*) .$$

B_X is then defined as $* \times_X E_X$.

Now, if there is a commutative diagram

(1.2)
$$\begin{array}{ccc} X & \xrightarrow{g} & Z \\ & {\scriptstyle h}\searrow & \downarrow{\scriptstyle h'} \\ & & R^+ \end{array}$$

with g a homomorphism, then there is an induced map $B_g : B_X \to B_Z$. Moreover, if g is an inclusion, then B_g is an inclusion and $Z \times_X E_X$ is the

---

[1] This is the geometric analogue of unique factorization in algebra.

restriction of E_Z to B_X. Thus the fiber in the map B_g is $Z \times_X E_X$. In particular, 395-396 of [18] shows that (1.2) is true for $X = J_1(Y)$ and $Y = J_{n+1}(Z)$, using the h_1, h_{n+1} constructed there. Passing to fiberings, we have proved

<u>Lemma</u> 1.3 <u>The fiber</u> F_n <u>in the natural map</u> $\Sigma Y \to \Omega^n\Sigma^{n+1}Y$ <u>is</u>

$$J_{n+1}(Y) \times_{J_1(Y)} E_{J_1(X)} .$$

This space admits a simple description as a CW complex.

<u>Corollary</u> 1.4 $C_\#(F_n) \cong C_\#(J_{n+1}(Y)) \otimes C_\#(\underset{\sim}{c}Y)$ <u>and</u> $\partial(a \otimes \underset{\sim}{c}(b)) = (-1)^{|a|} a \circ b \otimes 1 + (-1)^{|a|+1} a \otimes \underset{\sim}{c}(\partial b) + \partial a \otimes \underset{\sim}{c}(b)$.

There is an algebraic functor ([18], §7) which defines for any chain complex A an associated chain complex $F^n s^n(A)$. It gives $C_\#(J_n(Y))$ when applied to $C_\#(Y)$, and we have ([18], Theorem 7.2)

<u>Theorem</u> 1.5 <u>Let</u> A, A' <u>be chain complexes over</u> Z_p (for p a prime), <u>and suppose</u> $f : A \to A'$ <u>is an augmentation-preserving chain map inducing isomorphisms in homology</u>. <u>Then</u> $F^n s^n(f) : F^n s^n(A) \to F^n s^n(A')$ <u>also induces isomor- in homology</u>.

There are inclusions $\gamma_{i,j} : F^i s^i(A) \hookrightarrow F^{i+j} s^{i+j}(A)$ which, applied to $C_\#(Y)$, are induced from the inclusion $J_i(Y) \hookrightarrow J_{i+j}(Y)$, and we have

<u>Corollary</u> 1.6 <u>Let</u> $X = \Sigma Y$ <u>for</u> Y <u>a connected</u> CW <u>complex</u>. <u>Then</u> $H_*(F_n, Z_p)$ <u>depends only on</u> $H_*(X, Z_p)$. (Precisely, there are functors $\mathfrak{F}_p(n)$ for each p from the category of graded Abelian groups to graded Abelian groups, and $\mathfrak{F}_p(n)(H_*(\Sigma Y)) \cong H_*(F_n)$.)

Proof There is an injection $\gamma : H_*(Y,Z_p) \to C_\#(Y) \otimes Z_p$ inducing isomorphisms in homology. Hence we can form the algebraic object corresponding to the complex in 1.3, $G = F^{n+1}s^{n+1}(H_*(Y,Z_p) \otimes \underset{\sim}{c}(H_*(Y,Z_p))$ with boundary as in 1.3. γ extends to a chain map $\tilde{\gamma} : G \to C_\#(J_{n+1}(Y)) \otimes C_\#(\underset{\sim}{c}Y)$. Now, filtering both sides by the dimension in $\underset{\sim}{c}(Y)$, we obtain an algebraic Leray-Serre spectral sequence with $E^2 = H_*(J_{n+1}(Y)) \otimes H_*(\Sigma Y)$ in both cases. Moreover, $E_2(\tilde{\gamma})$ is evidently an isomorphism of E^2 terms. 1.6 now follows from the comparison theorem.

Remark 1.7 Suppose we consider the inclusion $F_n \to \Omega^L\Sigma^L F_n$, and study its fiber $F_{n,L}$. If $X = \Sigma^2 Y$, then the model 1.3 for F_n admits a natural description in terms of spaces $C_i^I \times (\Sigma Y)^i$ where C_i^I is a cell, and identifications are made over ∂C_i^I or when a coordinate in $(\Sigma Y)^i$ is $*$. There is a natural way to construct loops in these sets; namely, by using the suspension coordinates in $(\Sigma Y)^i$. Thus $(\Sigma Y)^i$ is an identification space of $I^i \times Y^i$, and one constructs paths in I^i starting at $(0\cdots 0)$ and ending at $(1,\ldots,1)$. A model for a sufficient number of paths is the Zilchgon $C(i-1)$, introduced in §4 of [18] (or in 0.18) Replacing I^i by $C(i-1)$ in each cell above, performing the appropriate identifications, and forming a universal construction, we obtain natural, minimal, and canonical models for ΩF_n, ..., $\Omega^{L+1}\Sigma^L F_n$, ...

These models also satisfy the property that the inclusions $\Omega^L\Sigma^{L-1}F_n \hookrightarrow \Omega^{L+1}\Sigma^L F_n$ ... are homomorphisms. Now we may apply 1.3 - 1.6 to show that $H_*(F_{n,L},Z_p)$ depends only on $H_*(Y,Z_p)$. Of course, if $X = \Sigma^3 Y$, we can iterate once more. In general, we have

Theorem 1.8 Define $F_{n_1\cdots n_k}(X)$ inductively as the fiber in the map

$$F_{n_1\cdots n_{k-1}}(X) \hookrightarrow \Omega^{n_k}\Sigma^{n_k}F_{n_1\cdots n_{k-1}}(X) .$$

Then for $X = \Sigma^k Y$, *it follows that* $H_*(F_{n_1 \cdots n_k}, Z_p)$ *depends only on* $H_*(X, Z_p)$.

We will have no further need of 1.7 and 1.8 except incidentally in the sequel; it is for this reason that the details are so skimpily sketched.

We now turn to the more limited observations which we can make about F_L when X is not a suspension.

Lemma 1.9 (Fiber Lemma) *Let* X , Y *be* n-1-*connected and locally finite* CW *complexes* $(n > 2)$. *Suppose* $f : X \to Y$ *satisfies* $f_* : H_t(X, Z_p) \to H_t(Y, Z_p)$ *is an isomorphism for* $t < 2n$ *and a monomorphism for* $t < 3n$. *Convert* f *into a fibering with fiber* F . *Then through dimensions* $3n-2$, F *is mod* p *weakly homotopy equivalent to* $\Omega(Y/X)$.

Proof F is 2-connected by our hypothesis. Thus, letting $\mathcal{C}$ be the class of finite groups having order prime to p , it is enough to show that there is a map $g : F \to \Omega(Y/X)$ so g_* induces a (mod $\mathcal{C}$)-isomorphism $H_*(F) \to H_*(\Omega(Y/X))$ in dimensions less than $3n-1$. But our first description gave F as $F = E_X{}^Y{}_*$, and g is defined as the evident projection

$$E_X{}^Y{}_* \to E_*{}^{Y/X}{}_* = \Omega(Y/X) \ .$$

Note that F and $\Omega(Y/X)$ are both 2n-2-connected. Thus the Serre spectral sequences for the fiberings

$$F \to X \to Y \ ,$$

$$\Omega(Y/X) \to P \to Y/X$$

are both exact sequences in dimensions $\leq 3n-2$. Moreover, both sequences split, and the fact that g_* is an isomorphism in this range follows.

Remark 1.10 1.9 can obviously be strengthened to give an actual homotopy equivalence in this range if f_* satisfies the hypothesis of 1.9 with the integers as coefficients.

In particular, we can now prove

Theorem 1.11 *Let* X *be a locally finite* $n-1$-*connected CW complex* $(n > 1)$. *Then through dimension* $3n-2$, *the fiber* F_n *in the inclusion*

$$X \hookrightarrow \Omega^n\Sigma^n(X)$$

is the space $\Omega(S^{n-1} \ltimes_T X \wedge X)$. (Here $S^{n-1} \ltimes_T X \wedge X$ is given as a quotient space of $S^{n-1} \times (X \wedge X)$ where (x,y,z) is identified with $(-x,y,z)$, and $(x,*)$ is set equal to $*$.)

Proof From [18], pp. 394-395 (or 0.24) we know $\Omega^n\Sigma^n X \simeq X \cup (I^{n-1} \times X \times X)/R$ through dimension $3n-1$ where R is a set of relations defined as follows:

1) $(t_1 \ldots \underset{\substack{\uparrow \\ j^{th} \text{ position}}}{\varepsilon} \ldots t_{n-1}, x, y) \sim (T^\varepsilon(t_1) \ldots T^\varepsilon(t_{j-1}),\ 0 \ldots 0) T^\varepsilon(x,y))$ where

$T(t) = 1-t$ and $T(x,y) = (y,x)$ if $\varepsilon = 0$ or 1.

2) $(t_1 \ldots t_{n-1}, *, y) \sim (t_1 \ldots t_{n-1}, y, *) \sim y$ for $*$, the base point of X.

Thus, through dimension $3n-2$, $\Omega^n\Sigma^n(X)/X \simeq (I^{n-1} \times X \wedge X)/R'$ where R' consists of relations of type (1), and $(t_1 \ldots t_{n-1}, *) \sim *$. Now, 1.11 follows from

Lemma 1.12 $(I^{n-1} \times X \wedge X)/R' \cong S^{n-1} \ltimes_T X \wedge X$.

Proof Embed $I^j \hookrightarrow I^{j+1}$ as the set of points $(\frac{1}{2}, t_1 \ldots t_j)$. This induces an embedding $(I^j \times X \wedge X)/R' \hookrightarrow (I^{j+1} \times X \wedge X)/R$. $(I^{j+1} \times X \wedge X)/R'$ can be given as the equivalence classes of points of the form $(t_1 \ldots t_{j+1}, x, y)$

with $t_1 \leq \frac{1}{2}$, with equivalence relations R' in $(I^j \times X \wedge X)$ together with $(0,t_2 \ldots t_{j+1},x,y) \sim (0,x,y)$, $(t_1,t_2,\ldots,t_{j+1},*) = *$. But $S^j \ltimes_T X \wedge X$ has an identical description in terms of the equitorial embedding $S^j \ltimes_T X \wedge X \to S^{j+1} \ltimes_T X \wedge X$. 1.11 follows by induction.

Theorem 1.13 *The equitorial inclusion* $S^{n-2} \hookrightarrow S^{n-1}$ *induces an inclusion* $\gamma_n : S^{n-1} \ltimes_T X \wedge X \to S^{n-1} \ltimes_T X \wedge X$, *and the diagram*

$$\begin{array}{ccc} F_{n-1} & \xrightarrow{k_n} & F_n \\ \downarrow & & \downarrow \\ \Omega(S^{n-2} \ltimes_T X \wedge X) & \xrightarrow{\gamma_n} & \Omega(S^{n-1} \ltimes_T X \wedge X) \end{array}$$

homotopy commutes. Here k_n is the map of fibers induced from the inclusion $h_{n-1} : \Omega^{n-1}\Sigma^{n-1}X \hookrightarrow \Omega^n\Sigma^n X$.

Proof This follows from the proof of 1.11 when we note that the inclusion

$$(X \cup I^{n-2} \times X \times X)/R \hookrightarrow (X \cup I^{n-1} \times X \times X)/R$$

defined on points by

$$(t_1,\ldots,t_{n-2},x,y) \mapsto (t_1,\ldots,t_{n-2},0,x,y)$$

induces a map H-homotopic to the map h_{n-1} (eg. see [18], §5).

Remark 1.14 The "little cubes" category of Boardman and Voit ([37]) provides an easy way of including the space $X \cup (S^{L-1} \ltimes_T X \times X)/R$ in $\Omega^L\Sigma^L X$, where R is the relation

$$(x,y,*) \sim (x,*,y) \sim y .$$

Specifically, the space $C^2(L)$ is defined as the set of all disjoint embeddings

$$I_1^L \cup I_2^L \to I_{(3)}^L$$

which are linear and take faces into sets parallel to the corresponding faces in $I_{(3)}$.

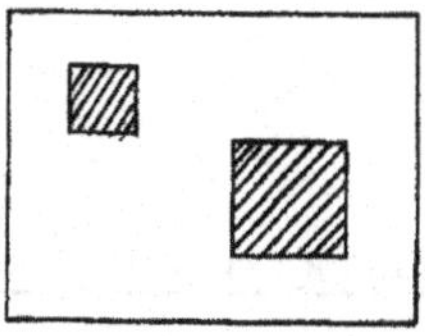

Z_2 acts (freely!) on $C^2(L)$ by interchanging I_1 and I_2 .

An arbitrary point x of $\Omega^L Y$ can be regarded as a map $f_x(I^L, \partial I^L) \to (Y,*)$. Given two maps f_x , f_y and a point z in $C^2(L)$, there is a map $(I^L, \partial) \to (Y,*)$ defined as

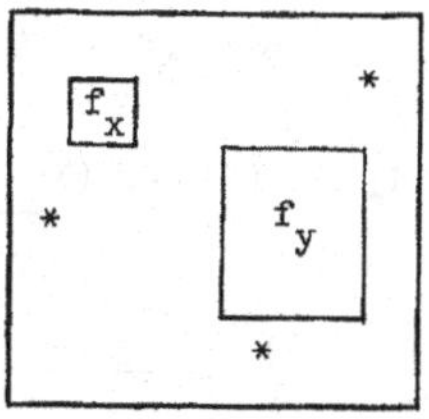

This provides a pairing $C^2(L) \ltimes_T \Omega^L(Y) \times \Omega(Y) \to \Omega^L(Y)$.

<u>Lemma</u> 1.14 (P. May) $C^2(L)$ <u>is equivariantly homotopic to</u> S^{L-1} <u>with the antipodal action</u>.

(Indeed, the equivariant inclusion $S^{L-1} \to C^2(L)$ takes x to the embedding of a cube of length $\frac{1}{4}$ with center at the point x , and the second with center at $-x$. It is now an easy geometric argument to prove 1.14.)

Now, using the pairing above in $\Omega^L\Sigma^L X$ and restricting to X, the desired inclusion of $X \cup (S^{L-1} \ltimes_T X \times X)$ in $\Omega^L\Sigma^L X$ is readily obtained.

§2. The Map $F_n(X) \to \Omega F_{n-1}(\Sigma X)$

Lemma 2.1 $\Sigma J_1(X) = \Sigma X \vee \Sigma(X \wedge X) \vee \Sigma(X \wedge X \wedge X) \vee \cdots$.

Proof This follows from the well-known ([38]) splittings

$$\Sigma X^n = \Sigma(\underbrace{X \wedge \cdots \wedge X}_{n}) \vee (n)\, \Sigma\, \underbrace{X \wedge \cdots \wedge X}_{n-1} \vee \cdots \vee \binom{n}{j}\, \Sigma\, \underbrace{X \wedge \cdots \wedge X}_{n-j} \vee \cdots,$$

writing $\Sigma J_1(X)$ as an identification space of $\bigcup_n \Sigma(X^n)$.

Corollary 2.2 There are H-maps $H_r : J_1(X) \to J_1(\underbrace{X \wedge \cdots \wedge X}_{r \text{ times}})$ so $(\otimes H_r)_* \; H_*J(X)) \to H_*(\otimes J_1\Sigma(X \wedge \cdots \wedge X))$ is injective.

H_r is called the rth Hopf-invariant map of X. Clearly, if an element α in $\pi_*(\Omega\Sigma X)$ comes from $\pi_*(X)$, then $H_{r*}(\alpha) = 0$ for all r.

Presumably there is a similar splitting for $\Sigma^n J_n(X)$. Thus it seems reasonable, in particular, to conjecture a splitting

$$\Sigma^2 J_2(X) = \Sigma^2 X \vee \Sigma^2 S^1 \ltimes_T X \wedge X \vee \Sigma^2 C(2) \ltimes_R (X \wedge X \wedge X) \vee \cdots.$$

Recent results of D. S. Kahn have shown that these splittings exist for $Q(X) = \lim_{n\to\infty} \Omega^n\Sigma^n X$; however, at present the splitting theorem for the J_n has not been proved.

Thus, we adopt an alternate "Hopf invariant" for the purposes of this section. The Hopf invariant of a class $\alpha \in \pi_m(\Sigma X)$ is defined by taking $\sigma^{-1}(\alpha) \in \pi_{m-1}(\Omega\Sigma X) = \pi_{m-1}(J_1(X))$, and projecting onto $\pi_{m-1}(J(X)/X)$. We denote its image by $H(\alpha)$, and, in the metastable range, the results of §1 show that $H(\alpha) = 0$ is both necessary and sufficient in order that α be in the image of σ .

In the metastable range, $H(\alpha) = H_2(\alpha)$ defined above.

Now we consider the problem of when an element $\beta \in \pi_m(\Sigma^L X)$ desuspends L-1 times but not L times.

Consider the diagram

$$\begin{array}{ccccc} F_L(X) & \to & X & \to & \Omega^L\Sigma^L(X) \\ \downarrow{\scriptstyle \tau_t} & & \downarrow{\scriptstyle \pi_t} & & \downarrow{\scriptstyle id} \\ \Omega^t F_{L-t}(\Sigma^t X) & \to & \Omega^t\Sigma^t X & \to & \Omega^L\Sigma^L X \end{array} .$$

Clearly, β does not desuspend L times if and only if $\partial\beta \neq 0$ in $\pi_*(F_L(X))$. On the other hand, it desuspends L-t times if and only if $\tau_{t*}(\partial\beta) = 0$. Thus the key step (outside of analyzing ∂ , which we defer for the moment) is to study the map τ_t . We can reduce this to the study of τ_1 since τ_t can clearly be decomposed as

$$F_L(X) \xrightarrow{\tau_1} \Omega F_{L-1}(\Sigma X) \xrightarrow{\Omega\tau_1} \Omega^2 F_{L-2}(\Sigma^2 X) \to \cdots \xrightarrow{\Omega^{t-1}(\tau_1)} \Omega^t F_{L-t}(\Sigma^t(X)) .$$

We have

<u>Theorem</u> 2.2 <u>Let</u> $f : \Sigma[S^{n-2} \ltimes_T (X \wedge X)] \to S^{n-2} \ltimes_T (\Sigma X \wedge \Sigma X)$ <u>be given on points by</u>

$$f(t, (x,y,z)) = (x\ (t,y),\ (t,z)) \ .$$

<u>Then the diagram</u>

$$\begin{array}{ccc} S^{n-2} \ltimes_T (X \wedge X) & \xrightarrow{\text{adj}(f)} & \Omega S^{n-2} \ltimes_T (\Sigma X \wedge \Sigma X) \\ & & \uparrow \\ F_n(X) & \xrightarrow{\tau_1} & \Omega F_{n-1}(\Sigma X) \end{array}$$

<u>homotopy commutes in the metastable range.</u>

<u>Proof</u> The map $\Omega(\Sigma X) \to \Omega[\Omega^{n-1}\Sigma^{n-1}(\Sigma X)]$ is given by the inclusion $J_1(X) \hookrightarrow J_n(X)$ ([18]). This inclusion satisfies the conditions of 1.9, so the diagram

$$\begin{array}{ccc} X & \to & J_n(X) \\ \downarrow & & \\ J_1(X) & \to & J_n(X) \end{array}$$

induces the inclusion of cofibers in dimensions less than $3n-1$,

$$S^{n-1} \ltimes_T X \wedge X \to (S^{n-1} \ltimes_T X \wedge X)/X \wedge X \ .$$

In the proof of Theorem 5 of [18], a map (adjoint to the identity) $\Sigma J_n(X) \to J_{n-1}(\Sigma X)$ is constructed. Precisely, there is a map $\varphi : I \times I \to I \times I$ defined by

$$\varphi(t,\tau) = \begin{cases} (2t,0), & t < \frac{1}{2}-\tau \\ (1-2\tau,\ 2(t+\tau-\frac{1}{2})), & \frac{1}{2}-\tau \le t \le 1-2\tau \\ (t,t), & 1-2\tau \le 2t \end{cases}$$

for $\tau < \frac{1}{2}$, and $\varphi(t, \frac{1}{2}+\tau) = T[\varphi(t,\tau)]$. This then defines a map

$$r : \Sigma[(X \cup I^{n-1} \times X^2)/R] \to (\Sigma X \cup I^{n-2} \times \Sigma X \times \Sigma X)/R$$

by $r(t,t_1,\ldots,t_{n-1},x,y) = \{t_2,\ldots,t_{n-1}(\varphi(t,t_1)x,y\}$ (identifying $I^2 \times X^2$ with $\Sigma X \times \Sigma X$) . In particular, $r(t,\varepsilon_1,t_2,\ldots,t_n,x,y) \to \Sigma X$. Thus, factoring X to a point, the induced map factors through

$$\Sigma[S^{n-1} \ltimes_T (X \wedge X)/X \wedge X] .$$

Finally, note that, for $t_1 = \frac{1}{2}$,

$$\varphi(t, \tfrac{1}{2}, t_2,\ldots,t_{n-1}, x, y) = \{t_2,\ldots,t_{n-1}(t,x, t,y)\} .$$

Thus 2.2 follows.

Actually, we have proved more than 2.2

<u>Corollary</u> 2.3 $\Sigma[S^{n-1} \ltimes_T (X \wedge X)/X \wedge X]$ <u>is homotopy-equivalent to</u> $S^{n-2} \ltimes_T \Sigma X \wedge \Sigma X$ <u>for</u> X <u>a connected</u> CW <u>complex</u>.

<u>Proof</u> Note that, using the cell decomposition of these spaces by the $[I^j \times X_r \times X_r]$, φ constructed above is cellular and induces an isomorphism of cellular chain complexes.

§3. The Cohomology of the F_n

We start by examining the cohomology of the $S^{n-1} \ltimes_T X \wedge X$. After that we give the structure of $H^*(F_n(\Sigma X))$.

There are maps

$$J : X \wedge X \to S^n \ltimes_T X \wedge X ,$$

$$K : S^n \ltimes_T X \wedge X \to \Sigma^n X \wedge X .$$

K is defined by identifying $S^{n-1} \ltimes_T X \wedge X$ to a point in $S^n \ltimes_T X \wedge X$.

Proposition 3.1 (a) J^* is surjective onto the invariant subalgebra under $(T)^*$ of $H^*(X \wedge X, Z_p)$ for p an odd prime. Moreover, kernel $J^* = \text{im}(K^*)$, and the following sequence is exact:

(3.2) $$H^*(\Sigma^n X \wedge X, Z_p) \xrightarrow{1+(-1)^n(\Sigma^n T)^*} H^*(\Sigma^n X \wedge X, Z_p) \to \text{im}(K^*) \to 0 .$$

(b) Mod (2) J^* is surjective as in (a), 3.2 is again exact, but there are additional elements $e^i \cup (\theta \otimes \theta)$ for $1 \le i \le n$ where $\theta \in \tilde{H}^*(X, Z_2)$, and these completely describe $H^*(\Gamma^n(X), Z_2)$.

Proof Consider the filtration of $\Gamma^n(X)$, $X \wedge X \hookrightarrow \Gamma^2(X) \hookrightarrow \dots \hookrightarrow \Gamma^n(X)$, obtained by embedding successive spheres equitorially. The resulting quotient spaces are the $\Sigma^i[X \wedge X]$. Moreover, the d_1 differential on $H^*(\Sigma^i X \wedge X, Z_p)$ is exactly $[1+(-1)^i(\Sigma^i T)^*]$.

Lemma 3.3 A chain complex for $S^n \ltimes_T X \wedge X$ is obtained as

$$W \otimes_T \widetilde{C \otimes C}$$

where C is any chain complex homotopy equivalent to $C_\#(X)$, and W is any free resolution of Z_2 (e.g. see [18], [27]). (This is immediate from the geometry.)

Now, to show $E_2 = E_\infty$ in our special sequence, note, for example,

$$\partial e_i \otimes (x \otimes x) = [(1+(-1)^i T)e_{i-1}] \otimes x \otimes x$$

if x is a cycle in $\tilde{C}$. But $(Te_{i-1}) \otimes x \otimes x = (-1)^{\dim x} e_{i-1} \otimes (x \otimes x)$ due to the action of T in $S^n \ltimes X \wedge X$. Hence

$$\partial e_i \otimes x \otimes x = \begin{cases} 0 , & i \not\equiv \dim (x)(2) \\ 2e_{i-1} \otimes x \otimes x , & i \equiv \dim x(2) \end{cases}$$

in $W \otimes_T \widetilde{C \otimes C}$. Thus cycles in E_2 are represented by cycles in $W \otimes_T \widetilde{C \otimes C}$, and 3.1 follows.

It is also fairly easy to verify that cup products (mod 2) are given by the formulae

$$(3.4) \qquad \begin{aligned} &[e^i \cup (\theta \otimes \theta)] \cup [e^j \cup (\tau \otimes \tau)] = e^{i+j} \cup (\theta\tau \otimes \theta\tau) , \\ &[e^i \cup (\theta \otimes \theta)] \cup \langle a,b \rangle = \begin{cases} 0 , & i > 0 , \\ \langle a\theta, b\theta \rangle , & i = 0 . \end{cases} \end{aligned}$$

Here, $\langle a,b \rangle$ is an appropriate choice of generator, so $J^*\langle a,b \rangle = a \otimes b + b \otimes a$, and $e^0 \cup \theta \otimes \theta$ is an element for which $J^*(e^0 \cup \theta \otimes \theta) = \theta \otimes \theta$. Indeed, 3.4 follows directly from:

<u>Lemma</u> 3.5 $\Delta : \Gamma^n(X) \to \Gamma^n(X) \times \Gamma^n(X)$ <u>admits a chain approximation</u>

$$\varphi : (W \otimes_{Z_2} \widetilde{C \otimes C}) \xrightarrow{\Delta_1 \otimes (\Delta_2 \otimes \Delta_2)} W \otimes W \otimes_{(Z_2 \times Z_2)} (\widetilde{C^2 \otimes C^2})$$
$$\xrightarrow{\text{shuff}} (W \otimes_{Z_2} \widetilde{C \otimes C}) \otimes (W \otimes_{Z_2} \widetilde{C \otimes C})$$

<u>where</u> Δ_1 <u>is a</u> $(T \otimes T, T)$ <u>equivariant diagonal map for</u> W , <u>and</u> Δ_2 <u>is any chain approximation to the diagonal in</u> X . (This is again immediate from the geometry; eg. see [27].)

We turn now to the question of higher Bocksteins. Mod (p) for p odd, these Bocksteins are determined by 3.1(a); however, their structure mod 2 is somewhat more involved.

Proposition 3.6 *Suppose* $\beta_i(a) = b$ *and:*

a) *dimension* a *is even; then* $\beta_j(a \otimes a) = 0, j < i$ *and* $\beta_i(a \otimes a) = \langle a,b \rangle$, *while* $Sq^1(b \otimes b) = e^1 \cup b \otimes b$;

b) *dimension* a *odd implies that* $Sq^1(a \otimes a) = e^1 \cup a \otimes a$

$$+ \begin{cases} 0, & i > 1 \\ \langle a,b \rangle, & i = 1, \end{cases}$$

while $\beta_{i+1}(\langle a,b \rangle + \theta e^1 \cup a \otimes a) = b \otimes b$;

c) *for* $i \geq 0$ *and* *dim* a *even,* $Sq^1(e^{2i+1} \cup (a \otimes a)) = e^{2i+2} \cup a \otimes a$, *and for* *dim* (a) *odd,* $Sq^1(e^{2i} \cup a \otimes a) = e^{2i+1} \cup a \otimes a$.

(The proof is a routine exercise using the explicit chain complex for $\Gamma^n(X)$ in 3.3; e.g., as in [20], [27].)

Finally, it remains to evaluate the action of $\mathcal{Q}(2)$ and $\mathcal{Q}(p)$ in $H^*(\Gamma^n(X))$. For p odd, this is immediate from 3.1 (modulo an extension problem, but that is handled in the next section; it turns out that the extension is trivial). Here is the result for $p = 2$.

Theorem 3.7 *Assume* $\theta \in H^n(X, Z_2)$. *Then:*

a) $Sq^i[e^k \cup (\theta \otimes \theta)] = \sum_{r,j} \binom{k}{r} \binom{n-j}{i-r-2j} e^{k+r-2j} \cup (Sq^j\theta) \otimes Sq^j\theta$,

b) $Sq^i(e^0 \cup \theta \otimes \theta) = \sum_{r<i} \langle Sq^r\theta, Sq^{i-r}\theta \rangle + \sum \binom{n-j}{i-2j} e^{i-2j} \cup Sq^j(\theta) \otimes Sq^j(\theta)$,

c) $Sq^i\langle a,b \rangle = \sum_{r=0}^{i} \langle Sq^r a, Sq^{i-r} b \rangle$,

at least modulo terms in *im* (K^*).

Proof The functor $\theta \to e^0 \cup \theta \otimes \theta$ is natural and representable, so it suffices to establish a , b for $\iota \otimes \iota$ where ι is the fundamental class of a $K(Z_2,n)$. Moreover, there is the map $\lambda : RP^\infty \times \ldots \times RP^\infty \to K(Z_2,n)$, taking ι to $e_1 \otimes \ldots \otimes e_n$ which induces a monomorphism in cohomology for dimensions $\leq 2n$. Reference to 3.1 (and the obvious naturality) shows that $\Gamma^n(\lambda) : \Gamma^n(RP^\infty \times \ldots \times RP^\infty) \to \Gamma^n(K(Z_2,n))$ also introduces a monomorphism in dimensions $\leq 4n$. Moreover, $\Gamma^n(\lambda)^*[\iota \otimes \iota] = (e^1 \otimes e^1) \cup \ldots \cup (e^n \otimes e^n)$. Thus we can apply the Cartan formula (noting by 3.4(b) that $Sq^1(e \otimes e) = \beta_1(e \otimes e) = e^1 \cup (e \otimes e) + \langle e^2, e \rangle$ and $Sq^2(e \otimes e) = (e \otimes e)^2 = e^2 \otimes e^2)$. Now a , b follow directly.

Finally, to prove (c), consider the map

$$S : RP^n \times K(Z_2,m) \times K(Z_2,\ell) \to \Gamma^n(K(m) \times K(\ell))$$

defined by $S(x,y,z) = \{x(y,z)(y,z)\}$. It is easy to show that $S^*\langle \iota_m, \iota_\ell \rangle = 0$, but $S^*[e^i \cup (Sq^I\iota_m \cup Sq^J\iota_\ell)^{(2)} = e^i \otimes (Sq^I\iota_m \cup Sq^J\iota_\ell)^2 + W$ is non-zero. Hence $Sq^i\langle \iota_m, \iota_\ell \rangle$ can only involve terms $\langle Sq^i\iota_m, Sq^{j-r}\iota_\ell \rangle$ and perhaps a term in $\mathrm{im}\,(K^*)$.

Finally, we will need the evaluation of the suspension map τ_1 (2.1) in cohomology.

Theorem 3.8 $(\sigma\tau_1)^*\langle a,b \rangle = 0$. $(\sigma\tau_1)^*(e^i \cup \sigma(a) \otimes \sigma(a)) = e^{i+1} \cup (a \otimes a)$.

The proof is direct from 2.3.

We now consider the structure of $H^*(F_n(\Sigma X))$. By use of the Eilenberg-Moore spectral sequence ([26], [30], [31]), there is a spectral sequence converging to $H_*(F_n(\Sigma X))$ having E^2 term $\mathrm{Tor}^*_{T[\tilde{H}_*(X,Z_p)]}(H_*(\Omega^{n+1}\Sigma^{n+1}(X), Z_p), Z_p)$. Here $T(A)$ is the tensor algebra on A , and $H^*(\Omega^{n+1}\Sigma^{n+1}(X), Z_p)$ is a module over $T[\tilde{H}_*(X,Z_p)]$ from the inclusion of H-spaces $J_1(X) \hookrightarrow J_{n+1}(X)$.

Moreover, the arguments of 1.6 show that $E^2 = E^\infty$. Thus, to calculate $H_*(F_n(\Sigma X))$, it suffices to calculate these Tor groups.

Note first that $H_*(J_{n+1}(X), Z_p) = P_n \otimes R$ where P is a polynomial algebra $P(\ldots\gamma_I(x)\ldots)$, with x running over $\tilde{H}_*(X,Z_p)$ and the γ_I over some basis for the universal loop homology operations, and R is $A(\tilde{H}_*(X,Z_p))$, the universal commutative algebra generated by $\tilde{H}_*(X,Z_p)$. The action of T on $P \otimes R$ is then obtained by projecting T on $1 \otimes R$.

Also, T is free; hence a resolution of T has the form $0 \to T \otimes s(\tilde{H}_*(X,Z_p)) \to T \to Z_p$. Now, tensoring with $P \otimes R$ and taking homology, we see that a basis for $\mathrm{Tor}^1_T(P \otimes R, Z_p)$ as a $P \otimes R$ module is given by the cycles

$$\langle a,b\rangle = as(b)-(-1)^{|a|+|b|}bs(a) \quad \text{for } a \neq b \text{ in } \tilde{H}_*(X,Z_p) .$$

We have thus calculated

Theorem 3.9 $H_*(F_n(\Sigma X), Z_p) \cong P_n \otimes L$ *where* L *is a module over* R *with generators* $\langle a,b\rangle$ *of degree* $|a|+|b|+1$, *where* a , b *are non-equal basis elements in* $\tilde{H}_*(X,Z_p)$. L *is completely determined as a module by the relations* $a\langle b,c\rangle-(-1)^{|a|\cdot|b|}b\langle a,c\rangle+(-1)^{|b|\cdot|c|+|a|\cdot|c|}c\langle b,a\rangle = 0$. *

We now turn to the cohomology structure of F_n . First note the fibering $\Omega^{n+1}\Sigma^{n+1}X \xrightarrow{J} F_n \to \Sigma X \to \Omega^n\Sigma^{n+1}X$.

---

* This proof was suggested by J. C. Moore.

Lemma 3.10 *There is a map* $\varphi : \Sigma F_n \to \Omega^n\Sigma^{n+1}X/\Sigma X$, *so the diagram*

$$\begin{array}{ccc} \Sigma\,\Omega^{n+1}\,\Sigma^{n+1}X & \xrightarrow{\text{adj }(1)} & \Omega^n\Sigma^{n+1}X \\ \Big\downarrow{\scriptstyle \Sigma(J)} & & \Big\downarrow{\scriptstyle \pi} \\ \Sigma\,F_n & \xrightarrow{\quad\varphi\quad} & \Omega^n\Sigma^{n+1}(X)/\Sigma X \end{array}$$

homotopy commutes. (π is the evident projection.)

Proof We can write $F_n = E_{\Sigma X}{}^{\Omega^n\Sigma^{n+1}(X)}{}_*$ (as in the remarks preceding 1.1), and take φ^* as the adjoint of the obvious map $F_n \to E_*{}^{\Omega^n\Sigma^{n+1}(X)/\Sigma(X)}{}_* = \Omega[\Omega^n\Sigma^{n+1}(X)/\Sigma(X)]$. The inclusion $J : \Omega^{n+1}\Sigma^{n+1} \to F_n$ is given in this notation by the natural inclusion $E_*{}^{\Omega^n\Sigma^{n+1}(X)}{}_* \hookrightarrow E_{\Sigma X}{}^{\Omega^n\Sigma^{n+1}(X)}{}_*$, and, by suspending and adjointing, 3.9 follows.

By tracing through the map φ , we find $\varphi_*\langle a,b\rangle = \pi_*(a\circ b)$ (the Pontrjagin Product) and $\varphi_*\alpha\circ\langle a,b\rangle = 0$. This gives us some information on passing to cohomology. However, to obtain more complete information, it is now necessary to use the action map

$$\Omega^{n+1}\Sigma^{n+1}(X) \times F_n \xrightarrow{\lambda} F_n$$

and our knowledge of the structure of this latter space as a module over the Steenrod algebra $\mathcal{A}(p)$ (e.g., see [21], [28]). Also note that λ_* is given by 3.8.

§4. The Structure Of Iterated Loop Spaces In The Metastable Range

We start with the following basic result:

Lemma 4.1 *Let* X *be the* n^{th} *loop space of* Y . *Then the fibering*

$$G_n \to \Omega^n\Sigma^n X \xrightarrow{\tau} X$$

has a cross-section. Hence, up to weak homotopy equivalence,

$$\Omega^n\Sigma^n X = G_n \times X \ .$$

Indeed,

$$\Omega G_n = F_n \ ,$$

the fiber in the inclusion

$$\pi : X \hookrightarrow \Omega^n\Sigma^n X \ .$$

Proof The cross-section of τ is exactly the inclusion $X \hookrightarrow \Omega^n\Sigma^n X$. Hence, since $\Omega^n\Sigma^n X$ is an H-space, there is a map

$$G_n \times X \xrightarrow{W} \Omega^n\Sigma^n X \ ,$$

and $\tau \circ W$ is projection on the first factor. Moreover, if we let G_n be given explicitly as $\Omega^n(H)$, where H is the fiber in the adjoint map $\Sigma^n X \to Y$, then the diagram

$$\begin{array}{ccc} G_n \times X & \xrightarrow{W} & \Omega^n\Sigma^n X \\ & {\scriptstyle \rho_2} \searrow \quad \swarrow {\scriptstyle \tau} & \\ & \vee & \end{array}$$

commutes up to reparametrization of paths. Now both maps are fiberings with fiber G_n . Thus, by the five-lemma, W is a weak equivalence.

Corollary 4.2 $H_*(X,Z_p) \cong H_*(\Omega^n\Sigma^n X,Z_p)//_{H_*(G_n,Z_p)}$ as Hopf algebras. In particular, if X is m-1-connected, then $H_*(X,Z_p)$ determines $H_*(G_n,Z_p)$ completely in dimensions $< 4m-1$, and G_n is 2m-1-connected. (Indeed, a Hopf algebra basis for $H_*(G_n,Z_p)$ may be given with generators $Q_I[\pi_*(x)] - \pi_*[Q_I(x)]$, $\pi_*(x_1) \circ \dots \circ \pi_*(x_n) - \pi_*(x_1 \circ \dots \circ x_n)$ where $x_1 \dots x_n$ run over a suitable basis for $H_*(X,Z_p)$, and the Q_I over a basis for the loop homology operations.)

Corollary 4.3 Suppose again that X is m-1-connected. Then in dimensions less than 3m-1 , G_n is homotopic to $S^{n-1} \ltimes_T X \wedge X$ for X, the homotopy type of a CW complex. Moreover, this equivalence is natural in the same range (from 1.11 and 4.1).

Corollary 4.4 Under the assumptions of 4.3, $H^i(X,Z_p)$ depends only on $H_*(Y,Z_p)$ for p odd, and on $H^*(Y,Z_2)$ as an $\mathcal{Q}(2)$-module for $i < 3m-1$.

Proof Since $G_n = \Omega^n H$, where H is the fiber in the map $\Sigma^n X \to Y$, 4.3 implies that $H \simeq \Sigma^n(S^{n-1} \ltimes_T X \wedge X)$ in dimensions less than $n+3m-1$. Also, in this range of dimensions, the Serre spectral sequence of this fibration becomes a long exact sequence

$$(4.5) \qquad \to H^i(H) \xrightarrow{\delta} H^{i+1}(Y) \xrightarrow{\sigma} H^{i+1}(\Sigma^n X) \xrightarrow{j} H^{i+1}(H) \to ,$$

and, to obtain $H^*(X,Z_p)$ for p a prime, it suffices to evaluate the map δ . For example, for $p = 2$, to evaluate $\delta(e^i \cup a \otimes a)$, consider the map

$$Y \xrightarrow{\sigma(a)} K(Z_2,\ell) .$$

By 4.3, this then induces a map of exact sequences

$$\begin{array}{ccccccc} \to H^i(H_{\ell,n}) & \to & H^{i+1}(K(Z_2,\ell)) & \to & H^{2+1}(\Sigma^n K(Z_2,n-\ell)) & \to & \\ \downarrow & & \downarrow & & \downarrow & & \\ \to H^i(H) & \to & H^{i+1}(Y) & \to & H^{i+1}(\Sigma^n X) & \to & . \end{array}$$

Moreover, $\Gamma^n(\sigma(\alpha))(e^i \cup \iota \otimes \iota) = e^i \cup \alpha \otimes \alpha$, and δ is determined by its behavior in the universal model. 4.4 follows.

In considering the proof of 4.4, it becomes clear that we need to know the map

$$\delta : H^i(H_{\ell,n}) \to H^{i+1}(K(Z_2,\ell),\ Z_2)$$

in order to determine the explicit form of the functor occurring in 4.4.

<u>Theorem</u> 4.6 $\delta\sigma^n[e^i \cup (Sq^I(\iota) \otimes Sq^I(\iota)] = Sq^{\ell + \deg(I) + i+1}(Sq^I(\iota))$, <u>while</u>

$$j^*\sigma^n(Sq^I(\iota) \cup Sq^J(\iota)) = \sigma^n\langle Sq^I(\iota),\ Sq^J(\iota)\rangle \qquad \text{(in 4.5)}.$$

<u>Proof</u> The second statement is obvious. To prove the first assertion, it suffices, by naturality and the known behavior (§§2,3) under suspension, to check that $\delta[\sigma(\iota_\ell \otimes \iota_\ell)] = Sq^{\ell+1}(\iota_{\ell+1}) = (\iota_{\ell+1})^2$ in the fibering $\Sigma K(Z_2,\ell) \to K(Z_2,\ell+1)$. Indeed, by the known results ([6]) on $H^*(K(Z_2,\ell+1))$, the kernel of σ^* in dimension $2\ell+2$ is exactly $(\iota)^2$. On the other hand, $H^{2\ell+1}(\Sigma(K(Z_2,\ell) \wedge K(Z_2,\ell)),\ Z_2) = Z_2$, and has generator $\sigma(\iota \otimes \iota)$. 4.6 follows.

<u>Remark</u> 4.7 The results 4.4, 4.6 give the structure of $H^*(\Omega^n Y, Z_p)$ completely in the metastable range as a module over Z_p . In particular, we have

Corollary 4.8 Suppose Y is m+n-1-connected, and $H^*(Y,Z_2)$ satisfies $Sq^I(a) = 0$ for excess $(I) > [\dim(a)-n]$. Then in dimensions less than $3m-1$, $H^*(\Omega^n Y,Z_2) = s^{-n}H^*(Y) \oplus H^*(G_n)$.

However, this splitting need not be valid over $\mathcal{Q}(2)$. For example, $CP^{12}{}_8$ certainly satisfies the hypothesis of 4.8 if $m = 5$. However, the secondary operation Φ_8, associated with the relation

$$Sq^1Sq^8 + Sq^2Sq^1(Sq^6) + Sq^8Sq^1 = 0 ,$$

is non-zero when evaluated on the bottom cell of $CP^{12}{}_8$. On the other hand, Φ_8 is universally zero on any 5 class. Thus, we must have $\Omega^{11}(\Phi_8(e_8))$ contained in the indeterminacy of (Φ_8). But this indeterminacy is zero in the part of $H^*(\Omega^{11}CP^{12}{}_8)$ coming from $\Omega^{11}(H^*(CP^{12}{}_8))$. Thus it must come from $H^*(G_{11},Z_2)$. In particular, there must be an element $\alpha \in H^*(G_{11},Z_2)$, so $Sq^2Sq^1(\alpha) = \Omega^{11}(\Phi_8(e_8))$.

§5. The Obstructions To Desuspension In The Metastable Range

We conclude the first part of this paper by considering a basic example. In the metastable range, we reduce the question of desuspension to the determination of when a certain map φ of known spaces is homotopy-trivial. Berstein and Ganea independently have obtained related results ([36]).

Theorem 5.1 Let X be n-1-connected, and have dimension less than $3n-2$. Then:

1) *if Y is the 2n-L-1 skeleton of X, there is a unique space Z, so* $\Sigma^L Z = Y$;

2) *X itself is an L-fold suspension if and only if a certain map*

$$\varphi : X/Y \to \Sigma^{L+1} S^{L-1} \ltimes_T Z \wedge Z$$

is homotopy-trivial;

3) *if* φ *is trivial, the number of distinct L-fold desuspensions of X is equivalent to the set of homotopy classes of maps*

$$[X/Y, \Sigma^L S^{2-1} \ltimes_T Z \wedge Z] .$$

(To avoid low dimensional complications, we also require L to be less than n-3 .)

Remark 5.2 In 5.1.3, the equivalence classes comprise: (a) distinct homotopy types of desuspensions, and (b) maps

$$h : W \to W$$

where $\Sigma^L W = X$, and where h is a homotopy equivalence $\simeq 1$ on the 2n-2L-1 skeleton of W .

Remark 5.3 By the dimensional restrictions, φ is actually a stable map since $S^{L-1} \ltimes_T Z \wedge Z$ is 2(n-L)-1-connected. Similarly, the set occurring in 5.1.3 is stable. Thus stable techniques are sufficient to determine them.

Proof $\Omega^L Y$ is a CW complex. The dimensional restrictions imply that the adjoint map $\Sigma^L \Omega^L Y \xrightarrow{\epsilon} Y$ induces an isomorphism in homology in dimensions $< 2n-L$. Now taking the associated cross-section of $\Sigma^L \Omega^L Y$, we can assume that Y is actually the 2n-L-1 dimensional skeleton of $\Sigma^L \Omega^L Y$. Thus Y is a suspension $\Sigma^L Z$. Moreover, the attaching map of every cell of Z is stable! Hence Z is indeed unique. This proves 5.1.1.

From 5.1.1, $X/Y = \Sigma^{L+1}W$ for a unique W . Moreover, there is a map $\tau : \Sigma^L W \to \Sigma^L Z$, so the following is a cofiber sequence:

(5.4) $$\Sigma^L W \xrightarrow{\tau} \Sigma^L Z \to X \to \Sigma^{L+1}W \to \dots .$$

Then X is a suspension $\Sigma^L(M)$ if and only if $\tau = \Sigma^L \tau'$ for some $\tau' : W \to Z$. Consider the diagram

(5.5) $$\begin{array}{ccc} \Omega^L\Sigma^L W & \xrightarrow{\Omega^L\tau} & \Omega^L\Sigma^L Z \\ \rho\uparrow & & \uparrow \\ W & \xrightarrow{\tau'} & Z . \end{array}$$

We assert that τ' exists if and only if the composite

(5.6) $$W \xrightarrow{\rho} \Omega^L\Sigma^L W \xrightarrow{\Omega^L\tau} \Omega^L\Sigma^L Z \xrightarrow{\pi} \Omega^L\Sigma^L Z/Z$$

is homotopy trivial. This follows from

Lemma 5.7 Let $\mathfrak{F}$ be the fiber in the map $\pi : \Omega^L\Sigma^L Z \to \Omega^L\Sigma^L Z/Z$. Then through dimension $3(n-L)-2$, $\mathfrak{F} \simeq Z$, and the inclusion $\mathfrak{F} \to \Omega^L\Sigma^L Z$ factors the inclusion $Z \to \Omega^L\Sigma^L Z$. (This is immediate from the fiber lemma, §1.)

On the other hand, from the proof of 1.11, it follows that, in the range of dimensions which concern us, $\Omega^L\Sigma^L Z/Z \simeq S^{L-1} \ltimes_T Z \wedge Z$, and this concludes the proof of 5.1.2 when we note (as in 5.3) that $\pi \circ \Omega^L\tau \circ \rho$ is a stable map, hence is homotopic to zero if and only if $\Sigma^{L+1}(\pi\circ\Omega^L\tau\circ\rho) = \varphi : X/Y \to \Sigma^{L+1}S^{L-1} \ltimes_T Z \wedge Z$ is.

To prove 5.1.3, note that different suspensions satisfying the relations imposed by 5.2 are given by different homotopy classes of liftings in (5.5). But by 5.7, these are given by maps of W into the fiber. $\Omega(\Omega^L\Sigma^L Z/Z) \simeq \Omega(S^{L-1} \ltimes_T Z \wedge Z)$ in our range. Dimensional considerations show that these are again stable, and 5.1.3 follows.

Remark 5.8 Combining 5.1 with 4.6 and the structure of $S^{L-1} \ltimes_T Z \wedge Z$, it is direct to calculate the first few obstructions explicitly in terms of higher order cohomology operations in X . For the first two obstructions, see [36] in particular.

Remark 5.9 Recent work of D. Anderson ([35]) makes it also possible to give analogues of 4.6 for certain exotic cohomology theories, eg. K-theory. This in turn makes it possible to carry through a program analogous to 5.8 in these theories as well. This remark will be considerably amplified in a forthcoming paper.

P A R T I I

§6. An Unstable Adams Spectral Sequence

In this section, we introduce a version of the Adams spectral sequence which gives information about the unstable homotopy of a space X. It is invariantly defined from E^2 on; however: (1) little is known about its convergence properties, and (2) in general, E^2 is not just a functor of $H^*(X,Z_p)$ over $\mathcal{A}(p)$ but actually depends on the space itself.

The construction we use is similar to the one given in [17]. However, due to the special cohomological properties of the spaces they considered, it there turns out that E^2 is a functor of $H^*(X,Z_p)$ over $\mathcal{A}(p)$.

In §8, we will show that, in the metastable range, E^2 is algebraically determined (explicitly) from $H^*(X,Z_2)$ over $\mathcal{A}(2)$. Thus we reduce many of the problems involved in metastable calculations to formal algebra and the determination of differentials in this sequence.

Definition 6.1 *An Adams* (p,q)-*resolution of a space* X *for* p *a prime and* q *a positive integer is a sequence of fiberings*

$$\begin{array}{ccccccc} X & \xleftarrow{\rho_1} & E_1 & \xleftarrow{\rho_2} & E_2 & \xleftarrow{\rho_3} & E_3 \\ \downarrow{\scriptstyle \pi_1} & & \downarrow{\scriptstyle \pi_2} & & \downarrow{\scriptstyle \pi_3} & & \downarrow{\scriptstyle \pi_4} \\ B_{H_1} & & B_{H_2} & & B_{H_3} & & B_{H_4} \end{array}$$

where:

i) E_i *is the fiber in the map* π_i ;

ii) B_{H_i} *is a generalized Eilenberg-MacLane space* $K(Z_p,\vec{n})$ *for the prime* p , *and*

iii) $\vec{n} = (n_1, \dots, n_i)$ *with each* $n_j \leq q$;

iv) $\rho_i^* : H^k(E_{i-1}, Z_p) \to H^k(E_i, Z_p)$ *is* 0 *for* $k \leq q$.

It is clear that, if X has the homotopy type of a CW complex, then (p,q)-resolutions of X exist for all (p,q) . They also satisfy the naturality properties:

Lemma 6.2 *Let* $f : X \to Y$ *be a map of* CW *complexes, and suppose given sequences*

a)

$$\begin{array}{ccccccccc} Y & \leftarrow & E_1' & \leftarrow & E_2' & \leftarrow & E_3' & \leftarrow \\ \downarrow & & \downarrow & & \downarrow & & \downarrow & \\ B_{H_1}' & & B_{H_2}' & & B_{H_3}' & & B_{H_4}' & \end{array}$$

where (a) *satisfies* (i), (ii) *of* 6.1, *and* (iii) *with* $q-1$ *in place of* q ;

b)

$$X \leftarrow E_1 \leftarrow E_2 \leftarrow E_3 \leftarrow$$

where (b) *satisfies all of* 6.1.

Then there are maps $f_i : E_i \to E_i'$, *so the diagram*

$$\begin{array}{ccccccccc} X & \leftarrow & E_1 & \leftarrow & E_2 & \leftarrow & E_3 & \leftarrow \\ \downarrow f & & \downarrow f_1 & & \downarrow f_2 & & \downarrow f_3 & \\ Y & \leftarrow & E_1' & \leftarrow & E_2' & \leftarrow & E_3' & \leftarrow \end{array}$$

commutes.

Lemma 6.3 *Under the assumptions of* 6.2, *suppose* $f \simeq g : X \to Y$ *and*

$$\{f_i : E_i \to E_i'\} ,$$

$$\{g_i : E_i \to E_i'\}$$

are given. Then there are homotopies

$$K_i : I \times E_i \to E'_{i-1} ,$$

so

i) $K_i(0,E_i) = \rho_i f_i$,

ii) $K_i(1,E_i) = \rho_i g_i$,

iii) *the diagram*

$$\begin{array}{ccccccccc}
I \times X & \xleftarrow{1\times\rho_1} & I \times E_1 & \xleftarrow{1\times\rho_2} & I \times E_2 & \xleftarrow{1\times\rho_3} & I \times E_3 & \longleftarrow & \cdots \\
{\scriptstyle K_0}\downarrow & \swarrow{\scriptstyle K_1} & & \swarrow{\scriptstyle K_2} & & \swarrow{\scriptstyle K_3} & & & \\
Y & \longleftarrow & E'_1 & \longleftarrow & E'_2 & \longleftarrow & \cdots & &
\end{array}$$

commutes.

As usual, taking the homotopy exact couple of the (p,q)-resolution in 6.1 gives a spectral sequence. Since $E_i \to E_{i-1} \to B_{H_i}$ is a fibering, it follows that $\pi_*(E_{i-1},E_i) \cong \pi_*(B_{H_i})$. Moreover, the d_1 differential is obtained by passing to homotopy in the composition

$$H_i \xrightarrow{j} E_i \xrightarrow{\pi_i} B_{H_{i+1}}$$

where $j : H_i \to E_i$ represents H_i as the fiber in the map ρ_i . By using 6.2, 6.3, we define our desired spectral sequence by passing to inverse limits over q when we note

Corollary 6.4 $E^2_{i,j}(X)[(p,q)]$ *depends only on* X *for* $i-j < q-1$, *and is isomorphic to* $E^2_{i,j}(X)$. *In particular,* $E^2_{i,j}(X)$ *depends only on* X *and not the resolutions used to take the inverse limit.*

Note that, when X is a (single) suspension, the convergence proof given in [34] carries over without change. (It depends only on 6.2 and the fact that it is possible to map a suspension onto a space Y_x, which satisfies 6.5(i) below for $x \in \pi_*(X)$, so the image of x in $\pi_*(Y_x)$ is non-zero.) Thus we have

Corollary 6.5(i) If X satisfies the condition that, for each j, there is an $r_j < \infty$ so $p^{r_j}\pi_j(X) \otimes Z_{(p)^\infty} = 0$, then the spectral sequence converges to $\pi_*(X) \otimes Z_{(p)^\infty}$.

(ii) If X is a suspension, then the spectral sequence always converges to $\pi_*(X) \otimes Z_{(p)^\infty}$.

For general X, the difficulty in extending 6.5 is in the elements of infinite order in $\pi_*(X)$. If X is simply connected, has the homotopy type of a locally finite CW complex, and is also an associative unitary H-space, then: (a) the only elements of $\pi_*(X)$ of infinite order are contained in Z-direct summands, and (b) have non-trivial images under the Hurewicz map ([43]).

Put another way, this says that the Postnikov invariants are all finite for X. An easy argument now shows:

Corollary 6.6 If X is a simply connected, locally finite unitary H-space, then the spectral sequence converges to $\pi_*(X) \otimes Z_{(p)^\infty}$.

6.5 and 6.6 establish convergence insofar as we need it. It is probable that more extensive results in this direction can be obtained from [44].

Example 6.7 Let $X = K(Z_{p^3},n)$. Then $B_{H_1} = K(Z_p,n) \times K(Z_p,n+1)$, and the k-invariants are (ι) and $\beta(\iota)$. It is easily verified that $E_1 = K(Z_{p^3},n)$, and the map ρ_{1*} is multiplication by p . Thus $E_2 = \dots = E_n = \dots = K(Z_{p^3},n)$, and ρ_{n*} is always multiplication by p . In particular, $E^1_{i,j} = E^2_{i,j} = 0$ unless $j-i = n,n+1$ when it is Z_p . The differentials are all d_3 's and are all non-trivial.

A map of spectral sequences $\sigma : E^*_{i,j}(X) \to E_{i,j+1}(\Sigma X)$ is defined from 6.2, 6.3, the map $\pi : X \hookrightarrow \Omega\Sigma X$, and the sequence

(6.8)
$$\begin{array}{ccccccc} \Omega\Sigma X & \xleftarrow{\Omega\rho_1} & \Omega E_1(\Sigma X) & \xleftarrow{\Omega\rho_2} & \Omega E_2(\Sigma X) & \longleftarrow \\ \downarrow & & \downarrow & & \downarrow & \\ \Omega B_{H_1} & & \Omega B_{H_2} & & \Omega B_{H_3} & \end{array}$$

associated to a (p,q)-resolution of ΣX .

<u>Corollary</u> 6.9 <u>If</u> X <u>is</u> n-1-<u>connected, the sequence of</u> (6.8) <u>satisfies</u> 6.1(iv) <u>in dimensions</u> $\leq 2n-2$. <u>Thus</u>

$$\sigma : E^*_{i,j}(X) \to E^*_{i,j+1}(\Sigma X)$$

<u>is an isomorphism in dimensions</u> $j-i \leq 2n-2$.

<u>Corollary</u> 6.10 <u>If</u> X <u>is</u> n-1-<u>connected, then</u>

$$E^2_{i,j}(X) \cong \operatorname{Ext}^{i,j}_{G(p)}(H^*(X), Z_p)$$

<u>for</u> $j-i \leq 2n-2$.

<u>Remark</u> 6.11 It is possible to generalize somewhat the above construction. Let $A \hookrightarrow H^*(X,Z_p)$ be an unstable sub-$G(p)$-module. Then we can resolve X by a sequence of fiberings

$$\begin{array}{ccccccc}
X & \overset{\rho_1}{\longleftarrow} & E_1 & \overset{\rho_2}{\longleftarrow} & E_2 & \overset{\rho_3}{\longleftarrow} & E_3 \longleftarrow \\
\downarrow \pi_1 & & \downarrow \pi_2 & & \downarrow \pi_3 & & \downarrow \pi_4 \\
B_{G(A)} & & B_{G_2} & & B_{G_3} & & B_{G_4}
\end{array} ,$$

so $\operatorname{im}(\pi_1)^*$ is exactly A, while the sequence $E_1 \leftarrow E_2 \leftarrow E_3 \leftarrow$ is a (p,q)-resolution of E_1. Analogs of 6.2 through 6.6 continue to hold. Thus we again have an Adams-type spectral sequence with invariantly-defined E^2-term. We denote it $E^2(X,A)$. Clearly, in the stable range, there is an exact sequence $\to \operatorname{Ext}^{i+1,*}_{G(p)}(A, Z_p) \to E^2_{i+1,*}(X,A) \to \operatorname{Ext}^{i,*}_{G(p)}(H^*(X)/A, Z_p) \xrightarrow{\partial} \operatorname{Ext}^{i+2,*}_{G(p)}(A,Z_p) \to \dots$, though, as we will see in the examples, ∂ may well be non-trivial.

7. The Loop Space Functor For Resolutions

Let X be m-1-connected, and suppose that

(7.1) $$\Sigma^n X \overset{\rho_1}{\longleftarrow} E_1 \overset{\rho_2}{\longleftarrow} E_2 \overset{\rho_3}{\longleftarrow} E_3 \longleftarrow \dots$$

is a $(2,q)$-resolution with $q < 3m+n$. In this section, we wish to study the behavior of (7.1) under the operation of taking loop spaces. It will appear that the sequence

(7.2) $$\Omega^n\Sigma^n X \overset{\Omega^n\rho_1}{\longleftarrow} \Omega^n E_1 \overset{\Omega^n\rho_2}{\longleftarrow} \Omega^n E_2 \longleftarrow \Omega^n E_3 \longleftarrow \dots$$

is not a $(2,q-n)$-resolution, as $(\Omega^n\rho_1)^*$ is not zero in general. However, this is the only point at which the sequence fails to satisfy the definition for a $(2,q-n)$-resolution. Moreover, we will be able to calculate exactly $\operatorname{im}(\Omega^n\rho_1)^*$.

We begin by proving

Theorem 7.3 *Under the above assumptions, the sequence*

$$\Omega^n E_1 \xleftarrow{\Omega^n \rho_2} \Omega^n E_2 \xleftarrow{\Omega^n \rho_3} \Omega^n E_3 \longleftarrow \cdots$$

is a (2,q-n)-*resolution of* $\Omega^n E_1$ *for* $q < 2n$.

Proof We proceed in two steps. First, consider the diagram of (vertical) fibrations

(7.4)
$$\begin{array}{ccccccc}
G_1 & \xleftarrow{f_1} & G_2 & \xleftarrow{f_2} & G_3 & \longleftarrow & \cdots \\
\downarrow \tau & & \downarrow \tau & & \downarrow \tau & & \\
\Sigma^n \Omega^n E_1 & \longleftarrow & \Sigma^n \Omega^n E_2 & \longleftarrow & \Sigma^n \Omega^n E_3 & \longleftarrow & \cdots \\
\downarrow \pi & & \downarrow \pi & & \downarrow \pi & & \\
E_1 & \longleftarrow & E_2 & \longleftarrow & E_3 & \longleftarrow & \cdots .
\end{array}$$

The key observation about (7.4) which we need is

Lemma 7.5 *Through dimension* $3m+n-2$, f_i^* *is the zero map in mod* p *cohomology*.

Proof From 4.3, $G_i = S^{n-1} \ltimes_T (\Omega^n E_i \wedge \Omega^n E_i)$ in our range. Moreover, by 3.1 and naturality, f_i^* is determined in our range by $(\Omega^n \rho_i)^*$ restricted to $H^r(\Omega^n E_i)$ with $r < 2m$. But by suspension, $H^r(\Omega^n E_i) \cong H^{n+r}(E_i)$ in this range. Hence $f_i^* = 0$, and 7.5 now follows.

Next we must consider the fibering

$$H_i \xrightarrow{j} E_i \to E_{i-1} .$$

Since $q < 2n$, we know that $j^* : H^r(E_i) \to H^r(H_i)$ is injective for $r < q$. Now consider the functor $\Sigma^n \Omega^n$ applied to the map j . We obtain the diagram of fibrations

(7.6)
$$\begin{array}{ccc} G_i' & \xrightarrow{\theta} & G_i \\ \downarrow & & \downarrow \tau \\ \Sigma^n\Omega^n H_i & \to & \Sigma^n\Omega^n E_i \\ \downarrow & & \downarrow \\ H_i & \xrightarrow{j} & E_i \end{array}$$

From 3.1, it now follows that θ_* is injective in our range. In particular, given $x \in H^r(E_i)$, we can suppose that $j^*(x) = \alpha$ and $\alpha \neq 0$; then the class $j^*(Sq^{\dim(\alpha)+k+1-n}(x)) = Sq^{\dim(\alpha)+k+1-n}(\alpha) \neq 0$. Thus $Sq^{\dim(\alpha)+k+1-n}(x) \neq 0$, provided, of course, dimension $(\alpha) < n+m$. From the exact cohomology sequence for the left-hand fibering in (7.6), we find

$$\delta(e^k \cup \alpha \otimes \alpha) = Sq^{\dim(\alpha)+k+1-n}(\alpha) .$$

Hence $\delta(e^k \cup x \otimes x) \neq 0$ in $H^*(E_i)$. Hence the only possible elements in $H^*(G_i)$ which are in the kernel of δ are the $\langle x,y\rangle^*$ with $x \neq y$ in $H^*(E_i)$. (Of course, x and y must each have dimension less than $n+m$.) Now note (as a consequence of 4.6) that

(7.7)
$$\langle x,y\rangle^* = \tau^*\sigma^n(\Omega^n(x) \cup \Omega^n(y)) ;$$

hence $\langle x,y\rangle^* \in \ker \delta$, and these elements give the entire kernel.

To complete the proof of 7.3, we suppose there is a $\lambda \in H^*(\Sigma^n\Omega^n E_i, Z_2)$ with $(\Sigma^n\Omega^n\rho_i)^*(\lambda) \neq 0$. In the diagram (7.4), it is certainly true that λ is not in the image of π^* .

Hence $\tau^*(\lambda) \neq 0$.

Hence ((7.6), (7.7)), it follows that

$$\lambda = \sigma^n(\Omega^n(x) \cup \Omega^n(y)) + \upsilon$$

for some υ in the image of π^* .

On the other hand, $(\Omega^n\rho_i)^*(\Omega^n(x) \cup \Omega^n(y)) = (\Omega^n\rho_i)^*(\Omega^n x) \cup (\Omega^n\rho_i)^*\Omega^n(y) = 0$
Also, trivially, $(\Sigma^n\Omega^n\rho_i)^*\pi^*(x) = 0$. But this implies

$$(\Sigma^n\Omega^n\rho_i)^*(\lambda) = 0 .$$

Hence $\lambda = 0$, and 7.3 is proved.

We now turn to the first map in (7.2), $\Omega^n\rho_1 : \Omega^n E_1 \to \Omega^n\Sigma^n X$. As in the proof of 7.3, we suspend and consider the diagram of (vertical) fibrations

(7.8)
$$\begin{array}{ccccc}
G & \xleftarrow{\ \mu\ } & G_1 & \xleftarrow{\ \theta\ } & G_1' \\
\downarrow{\scriptstyle\tau} & & \downarrow{\scriptstyle\tau_1} & & \downarrow \\
\Sigma^n(\Omega^n\Sigma^n X) & \xleftarrow{\ \Sigma^n\Omega^n\rho_1\ } & \Sigma^n\Omega^n E_1 & \longleftarrow & \Sigma^n\Omega^n H_1 \\
\downarrow{\scriptstyle\pi} & & \downarrow{\scriptstyle\pi_1} & & \downarrow \\
\Sigma^n X & \xleftarrow{\ \rho_1\ } & E_1 & \xleftarrow{\ j\ } & H_1
\end{array}$$

Note that π has a homotopy inverse. Hence, in the metastable range,

(7.9)
$$\Sigma^n\Omega^n\Sigma^n X \simeq \Sigma^n X \vee G .$$

Before proceeding further, we need to consider the bottom (horizontal) fibration in (7.8). For $\sigma^n(\alpha)$ in $H^*(\Sigma^n X, Z_2)$, let $\underset{\sim}{t}(\alpha)$ be any element in $H^*(H_1)$ which satisfies

(7.10)
$$\delta(\underset{\sim}{t}(\alpha)) = \sigma^n(\alpha)$$

in the Serre exact sequence of the fibration. Note that, for all $k \geq 0$, we have

(7.11)
$$Sq^{\dim(\alpha)+k+1}(\underset{\sim}{t}(\alpha)) \in \operatorname{im}(j^*) .$$

(In our range, this is due to exactness and the fact that $Sq^{\dim(\alpha)+1+k}(\alpha) \equiv 0$ for $k \geq 0$. For general $k \geq 0$, it follows from the Borel transgression theorem.) Hence, for each $k \geq 0$, we can choose

$$\beta_k(\alpha) \in H^*(E_1)$$

satisfying

$$j^*(\beta_k(\alpha)) = Sq^{\dim(\alpha)+k+1}(\underset{\sim}{t}(\alpha)) . \tag{7.12}$$

<u>Theorem</u> 7.13 <u>Under the splitting</u> (7.9), <u>let</u> $q_k(\alpha)$ <u>be the cohomology class corresponding to</u> $e^k \cup (\alpha \otimes \alpha)$ <u>in</u> $H^*(\Sigma^n\Omega^n\Sigma^n X)$. <u>Then for some choice of</u> $\beta_k(\alpha)$, <u>we have</u>

$$(\Sigma^n\Omega^n\rho_1)^* q_k(\alpha) = \beta_k(\alpha) .$$

<u>Proof</u> It suffices to verify 7.13 in the universal situation. The space which is universal for $\sigma^n(\alpha)$ is $\Sigma^n(K(Z_2, \dim(\alpha)))$. Thus, in place of the bottom line of (7.8), we consider the fibering

$$K(Z_2, n-1+\dim(\alpha)) \xrightarrow{\Phi} \mathcal{U} \to \Sigma^n(K(Z_2, \dim(\alpha))) . \tag{7.14}$$

Of course,

$$\underset{\sim}{t}(\sigma\iota_{\dim(\alpha)}) = \iota_{n+1+\dim(\alpha)} .$$

Now consider the fibering

$$\Omega^n(\mathcal{U}) \xrightarrow{\theta} \Omega^n\Sigma^n K(Z_2, \dim(\alpha)) \to K(Z_2, \dim(\alpha)) . \tag{7.15}$$

Clearly, $\Omega^n\mathcal{U}$ is $2\dim(\alpha)-1$-connected, and $\theta^*(\iota * \iota) = \nu$, the first non-zero element in $H^*(\mathcal{U})$. (Here, $\iota * \iota$ is the class dual to the Pontrjagin product.) By 4.3, $\Omega^n\mathcal{U} = S^{n-1} \ltimes_T K(Z_2, \dim(\alpha)) \wedge K(Z_2, \dim(\alpha))$ in the metastable range, and 7.13 follows by naturality under suspension. (Explicitly, the class corresponding to $e^k \cup \iota \otimes \iota$ in $H^*(U,Z_2)$ pulls back under φ^* in (7.14) to $Sq^{k+1+\dim(\alpha)}\iota$.)

Remark 7.15 Finally, it should be noted that 7.3 is valid whether or not the space Y with which we start is $\Sigma^n X$. Indeed, the only crucial condition is that Y be n+m-1-connected.

8. The Metastable Exact Sequence

In this section, we assume X is m-1-connected, and restrict our attention to the 3m-2 skeletons of all the spaces under consideration.

Consider the fibering $F_L \to X \to \Omega^L\Sigma^L X$; then, as we have observed, F_L has the homotopy type of

$$\Omega(S^{L-1} \ltimes_T X \wedge X)$$

in our range. Consequently, $H^*(F_L,Z_2)$ has a natural sub-$\mathcal{G}(2)$-module $A = \{\sigma^*\langle a,b\rangle \ ; \ a,b \in H^*(X)\}$.

A plays a fundamental role in the sequel. Before proceeding to our main result, 8.5, we illustrate why this must be so.

Lemma 8.1 Consider the inclusion

$$\theta : \Omega^{L+1}\Sigma^L X \to F \quad .$$

Then $\ker(\theta^*)$ is exactly A . (This follows immediately from the results of Part I (notably 3.10).)

Now suppose we start to construct (2,q)-resolutions. Observe that the composite

(8.2) $$X \to \Omega^L\Sigma^L X \to \Omega^L\Sigma^L X \to \Omega^L B_{H_1}$$

is onto in cohomology. Hence the first stage in the resolution of X can be assumed to be the fiber in the map of (8.2). The following diagram of (vertical and horizontal) fibrations is thus obtained.

(8.3) $$\begin{array}{ccccc} F_L & \xrightarrow{=} & F_L & & \\ \downarrow{\scriptstyle \bar{j}} & & \downarrow{\scriptstyle j} & & \\ E_1' & \longrightarrow & X & \longrightarrow & \Omega^L B_{H_1} \\ \downarrow{\scriptstyle \bar{\pi}} & & \downarrow{\scriptstyle \pi} & & \downarrow{\scriptstyle =} \\ \Omega^L E_1 & \xrightarrow{\Omega^L(\rho_1)} & \Omega^L\Sigma^L X & \longrightarrow & \Omega^L B_{H_1} \end{array}$$

(since E_1' can be regarded as the pull-back via $\Omega^L(\rho_1)$ of the fibering $\pi : X \to \Omega^L\Sigma^L X$) .

<u>Lemma</u> 8.4 (a) $$\operatorname{im}(\bar{j})^* = A ,$$

(b) $$H^*(E_1') \cong A \oplus B$$

<u>where</u> $B = H^*(\Omega^L(E_1))/\operatorname{im}\Omega^L(\rho_1)^*$.

<u>Proof</u> In our range, the Leray-Serre spectral sequence becomes a long exact sequence. Note that, in the sequence of the left-hand fibering in (8.3), $\delta\langle x,y\rangle = x \cup y$ and $\Omega^L(\rho_1)^*\, x \cup y = \Omega^L(\rho_1)^*(x) \cup \Omega^L(\rho_1)^*(y) = 0$. On the other hand, $\delta e^i \cup (x \otimes x) = (Q_i(x))^*$, and the results of §7 show that $\Omega^L(\rho_1)^*(Q_i(x))^* \neq 0$. Indeed, the $\Omega^L(\rho_1)^*(Q_i(x))^*$ span $\operatorname{im}\Omega^L(\rho_1)^*$. 8.4 follows.

We are now ready to state the main result of Part II.

<u>Theorem</u> 8.5 <u>Suppose</u> $L > 3m$. <u>Then there are maps of spectral sequences</u>

$$\partial^i{}_* : E^*_{i,j}(\Sigma^L X) \to E^*_{i,j-L-1}(F_L, A) ,$$

$$J^i{}_* : E^*_{i,j}(F_L, A) \to E^*_{i+1,j+1}(X) ,$$

<u>and the resulting sequences</u>

$$\xrightarrow{J_2} E_{i,j+1}{}^2(X) \xrightarrow{\sigma^L} E^2_{i,j+L+1}(\Sigma^L X) \xrightarrow{\partial_2} E^2_{i,j}(F_L, A) \xrightarrow{J_2} \cdots$$

<u>are exact for</u> $j-i \le 3m-2$, and converge to the generalized E.H.P. sequence.

<u>Remark</u> 8.6 The mechanism for determining $\partial^i{}_*$ will be made clear during the course of the proof.

<u>Proof</u> We begin by constructing the maps ∂_* , J_* .

Consider the first stage of an (F_L, A)-resolution of F_L :

$$(F_L)_1 \xrightarrow{\rho} F_L \xrightarrow{\varphi} B_{(A)}$$

where $\operatorname{im} \varphi^* = A$. From Lemma 8.1, there is no obstruction to extending θ to a map θ_1 so

$$\begin{array}{ccc} \Omega^{L+1}E_1 & \xrightarrow{\theta_1} & (F_L)_1 \\ {\scriptstyle \Omega^{L+1}\rho_1}\downarrow & & \downarrow{\scriptstyle \rho} \\ \Omega^{L+1}\Sigma^L X & \xrightarrow{\theta} & F_L \end{array}$$

commutes. The map θ_1 may now be continued to a map of resolutions. This gives ∂_* on passing to spectral sequences. Moreover, by 6.3, ∂_* is invariantly defined from E^2 on.

<u>Remark</u> 8.7 From 7.13, it follows that $\theta_1^{\;*}(e^i \cup x \otimes x) = (\Omega^{L+1}\rho_1)^*(Q_i(x_*))^*$. Thus the map $\theta_1^{\;*}$ is completely determined except on secondary elements associated to the resolution of A . We shall discuss this situation more carefully later in this section.

Now, consider the map

$$F_L \xrightarrow{j} X .$$

Since $\pi^* : H^*(\Omega^L\Sigma^L X) \to H^*(X)$ is surjective, it follows that $j^* = 0$. Let E_1' be the first stage in a resolution of X . It follows that j lifts to

$$\bar{j} : F_L \to E_1' .$$

From 8.4(a), $\operatorname{im}(\bar{j})^* = A$; thus if E_2' is the next stage in a resolution of X , it follows that $\bar{j}$ lifts to

$$\bar{j}_1 : (F_L)_1 \to E_2' .$$

We then continue $\bar{j}_1$ to a map of resolutions

(8.8)
$$\begin{array}{ccccccccccc}
\cdots \to & (F_L)_4 & \to & (F_L)_3 & \to & (F_L)_2 & \to & (F_2)_1 & \to & (F_L) & \\
 & \swarrow & & \swarrow & & \swarrow & & \swarrow{\scriptstyle \bar{j}} & & \downarrow{\scriptstyle j} & \\
\cdots \to & E_4' & \to & E_3' & \to & E_2' & \to & E_1' & \to & X & .
\end{array}$$

This defines J_* . Of course, J_* is not,as it stands, well-defined since it depends on the choice of lifting $\bar{j}$. However, anticipating the fact that the sequence 7.5 is exact, it is certain that the image of J_2 is invariant under the different choices of liftings.

In order to demonstrate this exactness, we construct a very special resolution of X , one which tries to be the direct sum of the resolution for $\Omega^L E_1$ and (F_L, A) . Only the first two steps, due to the influence of A , must be handled with special care.

To begin, take E_1' as given in (8.3). Consider the situation in (8.3). We have the fibering

$$\Omega^L E_2 \to \Omega^L E_1 \to \Omega^L B_{H_2} ,$$

and we can induce a fibering over $\Omega^L E_2$ from the fibering over $\Omega^L E_1$ on the left in (8.3). This gives the diagram of (vertical) fiberings

(8.9)
$$\begin{array}{ccccc}
F_L & \xrightarrow{=} & F_L & & \\
\downarrow & & \downarrow & & \\
\mathcal{E}_2 & \xrightarrow{\bar{\rho}} & E_1' & & \\
\downarrow & & \downarrow & & \\
\Omega^L E_2 & \xrightarrow{\Omega^L \rho_2} & \Omega^L E_1 & \longrightarrow & \Omega^L B_{H_2}
\end{array}$$

From 8.4(b), $\text{im}\,(\bar{\rho})^* = A$ since the map $(\Omega^L \rho_2)^* \equiv 0$ in our range. Thus we can kill A in $H^*(\mathcal{E}_2)$. This gives us the diagram

(8.10)

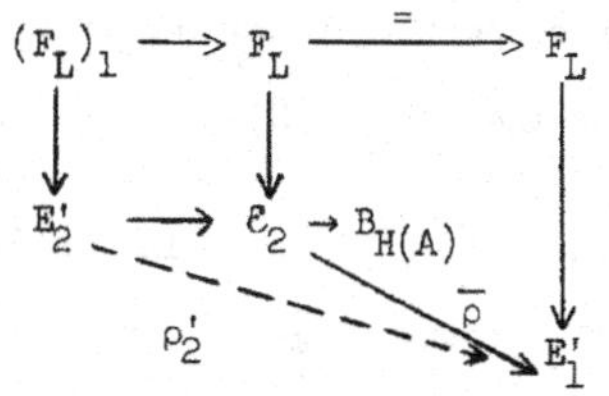

<u>Lemma</u> 8.11 <u>In (8.10), the composite map ρ_2' has as fiber a generalized Eilenberg-MacLane space.</u>

<u>Proof</u> Consider the diagram

(8.12)

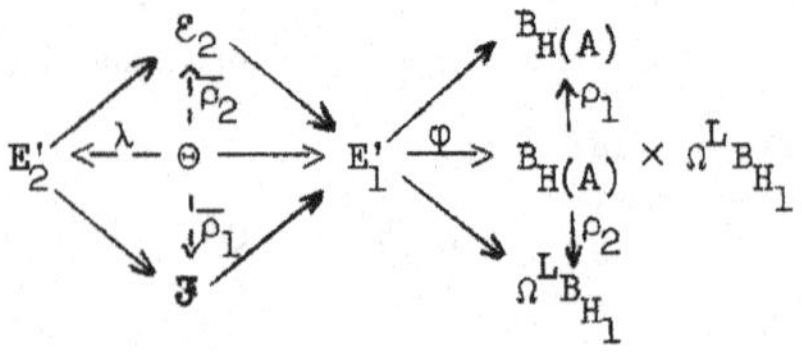

Here the 45° lines are fiberings, as well as the horizontal line starting with Θ . Clearly, there are liftings $\bar{\rho}_1$, $\bar{\rho}_2$ to make the triangles in (8.12) commute. But E_2' is universal for such pairs of maps. Hence there is a $\lambda : \Theta \to E_2'$ making 8.11 commutative in its entirety. Thus we have a map of fibrations

(8.13)
$$\begin{array}{ccccc} H(A) \times \Omega^L(H_1) & \to & \Theta & \to & E_1' \\ \lambda| \downarrow & & \downarrow \lambda & & \downarrow = \\ F & \longrightarrow & E_2' & \to & E_1' \end{array}$$

Note $\pi_*(F) \cong \pi_*(H(A) \times \Omega^L H_1)$, and reference to (8.13) shows that the isomorphism is actually induced by $\lambda|_*$. Θ , E_2' have the homotopy type of CW complexes. Thus, since the five-lemma shows $\lambda_* : \pi_*(\Theta) \to \pi_*(E_2')$ is an isomorphism, it follows that $\Theta \simeq E_2'$. 8.11 follows.

Again in view of 8.4(b), $(\rho_2')^* = 0$ in our range. Hence, from 8.11, it follows that ρ_2' is the second stage in our resolution.

To continue the process, consider the diagram

(8.14)

$$\begin{array}{ccccc}
(F_L)_1 & \xrightarrow{=} & (F_L)_1 & \longrightarrow & F_L \\
\downarrow & & \downarrow & & \downarrow \\
\mathcal{E}_3 & \longrightarrow & E_2' & \xrightarrow{\rho_2'} & E_1' \\
 & & {\scriptstyle\omega}\downarrow & & \\
 & & \mathcal{E}_2 & \longrightarrow & E_1' \\
\downarrow & & {\scriptstyle\pi}\downarrow & & \downarrow \\
\Omega^L(E_3) & \xrightarrow{\Omega^L\rho_3} & \Omega^L(E_2) & \longrightarrow & \Omega^L E_1
\end{array}$$

Here $\mathcal{E}_3$ is the fibering over E_2' induced by $(\pi\cdot\omega)$ from $(\Omega^L\rho_3)$. Since $(\Omega^L\rho_3)^* = 0$, it follows as in 8.4 that

(8.15) $$H^*(\mathcal{E}_3) \cong H^*((F_L)_1) \oplus H^*(\Omega^L E_3)$$

in our range. We construct E_3' by now killing $H^*((F_L)_1)$ in $H^*(\mathcal{E}_3)$. As in the proof of 8.11, we find that the composite

$$E_3' \to \mathcal{E}_3 \to E_2'$$

has as fiber a generalized Eilenberg-MacLane space. Moreover, by construction, the map is zero in cohomology.

This process evidently continues giving us a resolution of X, with fiber at each stage the product of a fiber in a resolution of F_L with a fiber in the resolution of $\Omega^L E_1$.

Passing to E^1-terms gives

(8.16) $$E^1_{i,j}(X) \cong E_{1,j+L}(\Sigma^L X) \oplus E^1_{i-1,j+1}(F_L, A) .$$

Now we need

Lemma 8.17 *The d_1 operator for the resolution given in (8.15) is the pair*

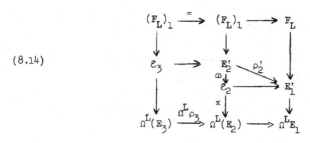

$$(\Omega^L \partial_1, \underset{\sim}{\partial}_1 \oplus \partial_{1(F_L,A)})$$

<u>where $\underset{\sim}{\partial}_1$ is the boundary map constructed in the first step of our proof from the resolution of $\Sigma^L X$ to that of (F_L,A)</u> .

Thus, algebraically, $E^2_{**}(X)$ is the homology of the mapping cone, and the exactness of 8.5 follows.

<u>Proof</u> We find immediately that the diagram

(8.18)

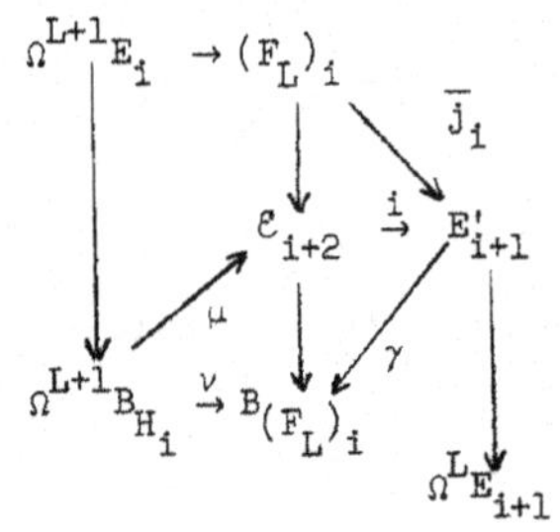

commutes. The fiber of the map

$$E'_{i+1} \to E'_i$$

is $\Omega^{L+1}B_{H_i} \times \Omega B_{(F_L)_{i-1}}$, and the map $i \circ u$ embeds $\Omega^{L+1}B_{H_i}$ as the first factor in this fiber. On the other hand, E'_{i+2} is the fiber in the map

$$\pi : E'_{i+1} \to B_{(F_L)_i} \times \Omega^L B_{H_{i+1}} ,$$

and $\gamma : E'_{i+1} \to B_{(F_L)_i}$ is exactly $p_1 \circ \pi$, the projection on the first factor.

The differential d_1 is determined by the map in homotopy in the composite

(8.19) $$\Omega^{L+1}B_{H_i} \times \Omega B_{(F_L)_{i-1}} \to E'_{i+1} \to \Omega^L B_{H_{i+1}} \times B_{(F_L)_i} ,$$

and now 8.17 follows.

This completes the proof of 8.5.

We now turn to the problem of determining the map ∂_* as explicitly as possible.

In 6.11, an exact sequence was exhibited which determines $E^2(F_L, A, Z_2)$ in terms of an exact sequence. Part of this sequence is a map

(8.20) $$\mu : E^2_{i,j}(F_L, A, Z_2) \to \mathrm{Ext}^{i-1,j+1}_{G(2)}(H^*(F_L)/A,\ Z_2) .$$

In terms of the map, we have

<u>Theorem</u> 8.21 <u>The composite</u>

$$\mu \circ \partial_2 : \mathrm{Ext}^{i,j}_{G(2)}(H^*(X),\ Z_2) \to \mathrm{Ext}^{i-1,j+2m-2}_{G(2)}(H^*(F_L)/A,\ Z_2)$$

<u>is algebraically determined, and commutes with the action of</u> $\mathrm{Ext}_{G(2)}(Z_2, Z_2)$ <u>on each of these modules</u>. (This is an immediate consequence of 8.7 and the fact that the map (8.20) is determined by the algebraic inclusion of $H^*(F_L)/A$ in $H^*((F_L)_1)$.)

<u>Remark</u> 8.22 In the special case that X is the sphere S^m , note that $A = \emptyset$ Thus (8.20) is an isomorphism, and we obtain as a corollary the main result of [13], on the existence of 8.5 for spheres.

<u>Remark</u> 8.23 In particular, this gives us many examples where ∂_2 is non-trivial. For example, if $X = S^7$, then, in the first non-trivial dimension, we have

$$\mathrm{Ext}^{1,8}{}_{\mathcal{G}(2)}(Z_2, Z_2)$$

has generator h_3, and $\mathrm{Ext}^{0,0}{}_{\mathcal{G}(2)}(H^*(P_7^{\infty}), Z_2) = Z_2$ with generator e_7, while

$$\partial_2(h_3) = e_7 .$$

9. Calculating The Groups $\mathrm{Ext}^{i,j}{}_{\mathcal{G}(2)}(H^*(F_L)/A, Z_2)$

In §8, we saw how the composite

$$(u \cdot \partial_*) : \mathrm{Ext}^{i,j}{}_{\mathcal{G}(2)}(H^*(X), Z_2) \to \mathrm{Ext}^{i-1,j-2L-2}{}_{\mathcal{G}(2)}(H^*(F_L)/A, Z_2)$$

is algebraically determined. In the context of the Adams spectral sequence approach to metastable homotopy theory, it replaces the classical ∂-invariant in the E.H.P. sequence.

In this section, we review the most effective techniques for calculating this Ext group for L greater than the connectivity of X .

$H^*(F_L)/A$ has generators $e^i \cup x \otimes x$ for $x \in H^*(X)$. There are two obvious filtrations for these elements:

(9.1) $$e^i \cup x \otimes x \in \mathfrak{F}_j \qquad \text{if } \dim x \geq j ,$$

(9.2) $$e^r \cup x \otimes x \in \mathcal{G}_i \qquad \text{if } r \geq i .$$

The results of §3 show that each of these $\mathfrak{F}_j$, $\mathcal{G}_i$ are closed under the action of the Steenrod algebra $\mathcal{G}(2)$. Hence they give rise to spectral sequences ${}_1E$, ${}_2E$ with

(9.3) $$ {}_1E^1_j \cong \mathrm{Ext}_{G(2)}(\mathfrak{F}_j/\mathfrak{F}_{j+1}, Z_2) , $$

(9.4) $$ {}_2E^1_j \cong \mathrm{Ext}_{G(2)}(G_i/G_{i+1}, Z_2) . $$

In order to identify these terms further, we note

Theorem 9.5 $$ \mathfrak{F}_j/\mathfrak{F}_{j+1} \cong H^*(P^\infty{}_j) \otimes H^j(X) . $$

Proof This is immediate from 3.1, or see [22, §2]. These groups have been tabulated in a range ([13]), and hence may be regarded as available for calculations.

Now, turning to the second filtration, let $S^2(H^*(X))$ be the T-invariant subalgebra of $H^*(X) \otimes H^*(X)$, and let

$$ B(H^*(X)) $$

be the quotient of $S^2(H^*(X))$ by the sub-$G(2)$-module image $(1+T)$. We evidently have

Theorem 9.6 The ${}_2E^1{}_{*\,k}$-term of the spectral sequence defined by filtration 2 is a copy of

$$ \mathrm{Ext}^{**}_{G(2)}(s^k B(H^*(X)), Z_2) . $$

There are cases, for example: when X has very few cells, when the first spectral is more convenient. However, when the structure of $H^*(X)$ is very intricate, but $\mathrm{Ext}_{G(2)}(H^*(X), Z_2)$ is completely known in a range, then the second spectral sequence is usually much easier to work with. (For example, in §11 we will study the case when $X = K(Z,m)$.)

In order to calculate the functor

$$\mathrm{Ext}_{\mathcal{G}(2)}(B(H^*(X)), Z_2) , \tag{9.7}$$

we proceed by exploiting the "doubling" homomorphism. Let $D : \mathcal{G}(2) \to \mathcal{G}(2)$ be the doubling map $(D(xy) = D(x)D(y),\ D(Sq^{2j}) = Sq^j)$. It is a map of algebras, and makes $\mathcal{G}(2)$ into an algebra over itself. We denote $\mathcal{G}(2)$ in this context as $\mathcal{G}(2)_D$. Similarly, we can "double" modules over $\mathcal{G}(2)$. Thus if M is a graded $\mathcal{G}(2)$-module, via the doubling map M becomes an ungraded $\mathcal{G}(2)$-module, $([Sq^i](m) = D(Sq^i)m)$.

<u>Lemma</u> 9.8 <u>There is a unique (graded)</u> $\mathcal{G}(2)$-<u>module</u> $D(M)$ <u>and ungraded isomorphism</u> $f : D(M) \to M$ <u>so that the diagram</u>

$$\begin{array}{ccc} \mathcal{G}(2) \otimes D(M) & \xrightarrow{D \otimes f} & \mathcal{G}(2) \otimes M \\ \downarrow & & \downarrow \mu \\ D(M) & \xrightarrow{f} & M \end{array}$$

<u>commutes</u>. (Obvious: $D(M)_{2i} \cong (M)_i$, $D(M)_{2i+1} = 0$.

We call $D(M)$ the double of M. Clearly, $B(H^*(X)) \cong D(H^*(X))$. We can expand the diagram of 9.5 as

$$\begin{array}{ccc} \mathcal{G}(2) \otimes D(\mathcal{G}) \otimes D(M) & \xrightarrow{u' \otimes 1} & D(\mathcal{G}) \otimes D(M) \\ D(u)\downarrow & & \downarrow D(u) \\ \mathcal{G}(2) \otimes D(M) & \longrightarrow & D(M) . \end{array}$$

In particular, $D(M)$ is actually a module over $D(\mathcal{G})$, and we can use the change of rings spectral sequence ([12]) converging to (9.7). Thus we have

Lemma 9.10 There is a spectral sequence converging to $\mathrm{Ext}_{\mathcal{a}(2)}(BH^*(X), Z_2)$ with E_2 term

$$\mathrm{Ext}_{\mathcal{a}(2)}(M, \widetilde{\mathrm{Ext}}_{\mathcal{a}(2)}(D\mathcal{a}(2), Z_2)) .$$

(Here the tilde over the internal Ext denotes the fact that this is a spectral sequence with twisted coefficients.)

However, 9.10 is still very useful since $\mathrm{Ext}_{\mathcal{a}(2)}(D\mathcal{a}(2), Z_2)$ has a very simple form; e.g., see [23, §2]. Specifically, we have

Theorem 9.11 $\mathrm{Ext}^{**}{}_{\mathcal{a}(2)}(D(\mathcal{a}), Z_2) \cong P(q_0, q_1, \ldots, q_i \ldots)$ where q_i has bidegree $(1, 2^{i+1}-1)$.

To illustrate methods, we outline the proof. The dual $\mathcal{a}^*$ of $\mathcal{a}(2)$ is a polynomial algebra on generators ξ_i of degree 2^i-1 . Clearly, the map $d : \xi_i \to \xi_i{}^2$ extendsto a map of Hopf algebras. Its dual is D ! Thus, dual to the sequence of Hopf algebras

(9.12) $$\mathcal{a}(2)^* \xrightarrow{d} \mathcal{a}(2)^* \to \mathcal{a}(2)^*//d\mathcal{a}(2)^* ,$$

there is the sequence

(9.13) $$(\mathrm{Ker}\ D) \to \mathcal{a}(2) \xrightarrow{\overline{D}} \mathcal{a}(2)\ //\mathrm{Ker}\ (D) \cong D(\mathcal{a}(2)) .$$

Of course, $\mathcal{a}(2)^*//d\mathcal{a}(2)^* \simeq E(\xi_1 \cdots \xi_i \cdots) \cong \mathrm{Ker}\ (D)$. Now the sequence (9.13) becomes an exact sequence of $\mathcal{a}(2)$-modules as

(9.14) $$0 \to \mathcal{a}(2) \cdot \overline{\mathrm{Ker}}\ (D) \to \mathcal{a}(2) \to D\mathcal{a}(2) \to 0 .$$

(9.14) gives rise to a long exact sequence of Ext groups, and, since $\mathcal{a}(2)$ is $\mathcal{a}(2)$-free, this implies

$$\mathrm{Ext}^{i}_{\mathcal{G}(2)}(D\mathcal{G}(2),\ Z_2) \cong \mathrm{Ext}^{i-1}_{\mathcal{G}(2)}(\mathcal{G}(2)\cdot\overline{\mathrm{Ker}}\ (D),\ Z_2)\ .$$

On the other hand, $\mathrm{Ext}^{i-1}_{\mathcal{G}(2)}(\mathcal{G}(2)\cdot\overline{B},\ Z_2) \cong \mathrm{Ext}^{i}_{B}(Z_2, Z_2)$ for any sub-Hopf algebra of $\mathcal{G}(2)$ ([23, §2]). Thus the result follows from the routine calculation of the Ext groups for a graded exterior algebra.

Remark 9.15 The difficulty with using 9.10 and 9.11 for the spectral sequence (9.4) is that the twisting in 9.10 is almost always non-trivial.

Example 9.16 In case $X = K(Z_2, m)$, we find the spectral sequence (9.4) by far the more convenient. Its E^1_r term in our range is a copy of $P(q_0 \cdots)$; however, the differentials are very involved. Considerable work on this case has been done by J. Harper ([10], [11]).

PART III

APPLICATIONS AND EXAMPLES

10. Calculations Of The Stable Homotopy Of $K(\pi,n)$'s

In [4], [5], and [14], methods were given for studying the stable homotopy groups $\pi^s_*(K(\pi,n))$. These techniques give only fragmentary information in actual practice. More recently, work of Browder and Brown ([38], [39]) and Brumfiel, Madsen, and the author ([40]) has shown the importance of these groups for problems relating to the classification of manifolds. Also, and rather tautologically, these groups give the classification of higher order one-variable cohomology operations which vanish for dimensional reasons, [3], [8], [10], [11] universally on appropriate cohomology classes.

In this section, we apply the methods of Sections 4 and 9 to the problem, obtaining somewhat more efficient tools for the calculations. We begin by considering the first few groups when $\pi = Z_2$ or Q/Z , the cases of interest in [38], [39], [40]. We then continue in the next section by giving more extensive calculations for the case $\pi = Z$.

In all cases, we begin by considering the fibering

$$G_L \to \Sigma^L K(\pi,n) \to K(\pi,n+L) \ . \tag{10.1}$$

This is of the type considered in 4.1; hence in our range

$$G_L = \Sigma^L S^{L-1} \ltimes_T (K(\pi,n) \wedge K(\pi,n)) \ .$$

Moreover, since $\pi_i(K(\pi,n+L)) = 0$ for $i > n+L$, it follows from the homotopy exact sequence of (10.1) that $\pi_i^s(K(\pi,n)) \cong \pi_i^s(G_L)$ in our range.

Thus to compute $\pi_i{}^S K(Z_2,n)$, we must calculate the homotopy of

$$S^{L-1} \ltimes_T K(Z_2,n) \wedge K(Z_2,n) .$$

In the first few dimensions, this has cells as follows:

$$\begin{array}{lllll} e_3 \cup \iota \otimes \iota & e_1 \cup Sq^1\iota \otimes Sq^1\iota & \langle Sq^1\iota, Sq^2\iota\rangle & \langle \iota, Sq^2Sq^1\iota\rangle & \langle \iota, Sq^3\iota\rangle \\ e_2 \cup \iota \otimes \iota & Sq^1\iota \otimes Sq^1\iota & \langle \iota, Sq^2\iota\rangle & & \\ e_1 \cup \iota \otimes \iota & \langle \iota, Sq^1\iota\rangle & & & \\ \iota \otimes \iota . & & & & \end{array} \tag{10.2}$$

The action of the Steenrod algebra $\mathfrak{A}(2)$ on the elements enumerated above depends, of course, on n . Thus for $n \equiv 2(4)$, we find from 3.7 that

$$Sq^1(\iota \otimes \iota) = \langle \iota, Sq^1\iota\rangle ,$$

$$Sq^2(\iota \otimes \iota) = Sq^1\iota \otimes Sq^1\iota + \langle \iota, Sq^2\iota\rangle + e_2 \cup \iota \otimes \iota$$

$$Sq^3\iota \otimes \iota = \langle Sq^2\iota, Sq^1\iota\rangle + \langle Sq^3\iota, \iota\rangle ,$$

$$Sq^1e_1 \otimes \iota \otimes \iota = e_2 \cup \iota \otimes \iota ,$$

$$Sq^2e_1 \otimes \iota \otimes \iota = e_3 \cup \iota \otimes \iota + e_1 \cup Sq^1\iota \otimes Sq^1\iota ,$$

$$Sq^1(Sq^1\iota \otimes Sq^1\iota) = e_1 \cup Sq^1\iota \otimes Sq^1\iota .$$

In this range, there are at least four non-trivial Z_2 's in homotopy; the first dual to $\iota \otimes \iota$, the second dual to $e_1 \cup \iota \otimes \iota$, the third dual to $Sq^1\iota \otimes Sq^1\iota$, and the fourth dual to $\langle Sq^1\iota, Sq^2\iota\rangle$. Applying 3.7 in the remaining cases, we obtain the table

	i	0	1	2
$n \equiv 0(2)$	$\pi^s_{2n+i}(K(Z_2,n))$	Z_2	Z_2	Z_2
	generator	$(\iota \otimes \iota)_*$	$(e_1 \cup \iota \otimes \iota)_*$	$(Sq^1\iota \otimes Sq^1\iota)_*$
$n \equiv 1(2)$	$\pi^s_{2n+i}(K(Z_2,n))$	Z_2	Z_4	Z_2
	generator	$(\iota \otimes \iota)_*$	$(e_1 \cup \iota \otimes \iota) + \langle \iota, Sq^1\iota \rangle)_*$	$(e_2 \cup \iota \otimes \iota)_*$

Now we consider the group Q/Z. Suprisingly, this involves us in some algebraic problems. The first is a suitable description of Q/Z. To this end, let $Z \quad = \varinjlim (Z_{p^i})$ for each prime p.

Lemma 10.4 $Q/Z \cong \oplus (Z \quad)$ *as* p *runs over all primes.*

Proof Injections $\theta_p : Z_{(p)^\infty} \to Q/Z$ are defined as the direct limits of maps

$$\rho_i : Z_{p^i} \to Q/Z$$

where $\rho_i(n) = n/p^i$. These in turn induce a morphism

$$\theta : \bigoplus \hat{Z}_{(p)^\infty} \to Q/Z$$

given on elements by

$$\theta(n_p \cdots n_q) = \theta_p(n_p) + \ldots + \theta_q(n_q) .$$

To see that $\ker(\theta) = 0$, note that

$$n_p/p^i + \ldots + n_q/q^j \in Z$$

if and only if $n_p(\cdots q^j) + \ldots + (p^i \cdots)n_q \equiv 0(p^i \cdots q^j)$. In particular,

$$\begin{aligned} n_p(\cdots q^j) + \ldots + (p^i \cdots)n_q &\equiv 0(p^i) \\ &\vdots \\ &\equiv 0(q^j) , \end{aligned}$$

and these congruences imply $n_p \equiv 0(p^i)$, ... , $n_q \equiv 0(q^j)$. Thus $\ker(\theta) = 0$. To see the converse, suppose the fraction $\frac{n}{2^i 3^j \ldots q^k}$ is given in Q/Z . Since the ideals $(p^j) + (q^k) = Z$ in the integers, for p a prime and q relatively prime to p , there are integers m_2 , n_2 so

$$n_2 2^i + m_2(3^j \cdots q^k) = n .$$

Similarly, there are integers m_3 , n_3 so

$$m_3(5^s \cdots q^k) + n_3 3^j = n_2 ,$$

etc. Hence we can write

$$n/2^i \cdots q^k = m_2/2^i + \ldots + m_q/q^k ,$$

and this shows that θ is onto. 10.4 follows.

Now note that the inclusion $Z_{p^i} \hookrightarrow Z_{p^{i+1}}$ induces an inclusion $K(Z_{p^i},n) \hookrightarrow K(Z_{p^{i+1}},n)$ (e.g., see the models for $K(\pi,n)$ introduced in [19]). Clearly, we have

<u>Lemma</u> 10.5 $K(Z_{(p)^\infty},n)$ $\varinjlim$ (weak) $K(Z_{p^i},n)$ <u>under these inclusions</u>. (Here, the topology in $K(Z_{(p)^\infty},n)$ is given by specifying its compact sets [which are the sets X contained in $K(Z_{p^i},n)$ for some i , so $X \cap K(Z_{p^i},n)$ is compact], and letting Y be open if and only if $Y \cap X$ is open for each compact set X .)

<u>Corollary</u> 10.6 $K(Q/Z,n) = \prod^{W}_{p \text{ prime}} K(Z_{(p)^\infty},n)$ (weak limit of the finite Cartesian products).

Corollary 10.7 Consider the fibering

$$G_L \to \Sigma^L K(Q/Z,n) \to K(Q/Z,n+L) \ .$$

Then $G_L = \prod_{p \text{ prime}} {}^W \varinjlim (\text{weak})\, \Sigma^L[S^{L-1} \ltimes_T K(Z_{p^i},n) \wedge K(Z_{p^i},n)]$ in our range.

Thus we can calculate the stable homotopy groups of $K(Q/Z,n)$ from the observation

10.8 $\pi_*(\varinjlim (\text{weak})\ X_i) \cong \varinjlim \pi_*(X_i)$.

Using 3.1 and 3.7, we now find

Theorem 10.9 (a) $\pi_{2n}{}^s(K(Q/Z,n)) = 0$,

$$\text{(b)}\quad \pi^s{}_{2n+1}(K(Q/Z,n)) \cong \begin{cases} Q/2Z \cong Q/Z\ , & n \text{ odd,} \\ 0\ , & n \text{ even.} \end{cases}$$

Proof $\pi_{2n+L}(G_L(K(Z_{p^i},n)) = \begin{cases} Z_{p^i}\ , & n \text{ even} \\ 0\ , & n \text{ odd},\ p \neq 2 \\ Z_2\ , & n \text{ odd},\ p = 2\ . \end{cases}$

In homology, $(\iota_{p^i})_* \to p(\iota_{p^{i+1}})_*$ under our injection. Hence $(\iota_{p^i})_* \otimes (\iota_{p^i})_* \to p^2((\iota_{p^{i+1}})_*)$, and the kernel of the map is generated by

$$p^{i-1}(\iota_{p^i})_* \otimes (\iota_{p^i})_* \ .$$

Moreover, the image of $p^{i-2}[(\iota_{p^i})_* \otimes (\iota_{p^i})_*]$ is in the kernel of the next map, and, in general, $p^{i-j}[(\iota_{p^i})_* \otimes (\iota_{p^i})_*]$ is in the kernel of the jth iterate. Thus the direct limit is indeed zero, and 10.9(a) follows.

To prove (b), note that

$$(10.10) \qquad \pi_{2n+1+L}(G_L(K(Z_{p^i},n))) = \begin{cases} Z_{p^i} , & n \text{ odd}, \quad p \neq 2 \\ Z_{2^{i+1}} , & n \text{ odd}, \quad p = 2 \\ 0 , & n \text{ even}, \quad p \neq 2 \\ Z_2 , & n \text{ even}, \quad p = 2 . \end{cases}$$

Moreover, from examining the action in homology of the Hurewicz images of the generators, we conclude that, in each case, the inclusion of the one in the next is given by multiplication by p . Hence

$$\varinjlim (Z_4 \subset , Z_8 \subset \ldots) = Z_{(2)^\infty}$$

is the two-local component of the limit. Clearly, there is a surjection of this $Z_{(2^\infty)}$ on $Z_{(2^\infty)}$ with kernel Z_2 . Thus it is natural to embed the resulting group in $Q/2Z$. This completes the proof of (b).

Remark 10.11 Theorem 10.9 is exactly the homotopy theoretic result which is needed to do simply-connected surgery ([41]), and also plays a key role in [40]. The results of Section 2 also allow us to know exactly at which suspension the various homotopy elements first appear. For example, the reader is asked to verify that, for n odd, a Q/Z appears in $\pi_{2n+1}(K(Q/Z,n) \wedge KQ/Z,n)$, and hence in $\pi_{2n+2}(\Sigma(KQ/Z,n))$. However, while the map

$$\pi_{2n+2}(\Sigma(K(Q/Z,n))) \to \pi_{2n+1}{}^{s}(K(Q/Z,n))$$

is an isomorphism, the approximating maps

$$Z_{2^i} \to Z_{2^{i+1}}$$

for $K(Z_{2^i},n)$ always have cokernel Z_2 .

11. Some Calculations Of The Stable Homotopy Groups For The $K(Z,n)$

In this section, we apply the techniques of §9 to study $\pi^S_*(K(Z,n))$ in some special cases. The calculational results illustrate the structure of the exact sequence 6.11.

(10.1) shows that we need study only $\pi_*(S^{L-1} \ltimes_T K(Z,n) \wedge K(Z,n))$ for $L > n$. To do this, we shall calculate the groups

$$\text{(11.1)} \qquad \begin{gathered} \mathrm{Ext}_{G(2)}^{**}(H^*(\Gamma_L K(Z,n)),\ Z_2)\ , \\ \mathrm{Ext}_{G(2)}^{**}(H^*(\Gamma_L),\ A,\ Z_2) \end{gathered}$$

in three steps. First, we calculate $\mathrm{Ext}_{G(2)}^{**}(H^*(G_L)/A,\ Z_2)$; then we calculate $\mathrm{Ext}_{G(2)}^{**}(A,Z_2)$; finally, we put them together to obtain the desired groups (11.1).

We begin with the calculation of $\mathrm{Ext}_{G(2)}^{**}(H^*(G_L)/A,\ Z_2)$, using the second spectral sequence in §9.

<u>Lemma</u> 11.2
$$\mathrm{Ext}^{**}{}_{G(2)}(D[H^*(K(Z,n))],\ Z_2) \cong$$
$$\cong \mathrm{Ext}^{**}{}_{G_1}(Z_2,Z_2) \otimes \mathrm{Ext}^{**}{}_{\Lambda(Q_3Q_4\cdots)}(Z_2,Z_2)$$

<u>in total degrees less than</u> $4n+1$. <u>Here</u> G_1 <u>is the subalgebra of</u> $G(2)$ <u>generated by</u> Sq^1 <u>and</u> Sq^2 .

<u>Proof</u> We imitate the main idea in the proof of 9.11. The map u $H^*(K(Z,n))$ is onto in dimensions less than or equal to $2n$ and has kernel $G(2)Sq^1$. Thus, the kernel of the doubling homomorphism

$$G(2) \xrightarrow{D} DG(2) \xrightarrow{D(u)} DH^*(K(Z,n))$$

is $D^{-1}(G(2)Sq^1) = \text{Ker}(D) \oplus G(2)(Sq^2)$, but this is $G(2)\overline{E}(Sq^1, Q_2, Q_3\cdots) \oplus G(2)(Sq^2)$, but this ideal is clearly isomorphic to

$$G(2)\cdot\overline{(G_1 \otimes E(Q_3Q_4\cdots))} ,$$

and 11.2 follows.

Now we recall the well-known

<u>Lemma</u> 11.3 $\text{Ext}^{**}_{G_1}(Z_2,Z_2) = P(h_0,h_1Q,P)/R$ <u>where</u> R <u>is the set of relations</u>

$$Q^2 = h_0^2P ,$$
$$h_0h_1 = 0 ,$$
$$h_1^3 = 0 ,$$
$$h_1Q = 0 .$$

<u>Here</u> h_0 <u>has bidegree</u> (1,1) , h_1 <u>has bidegree</u> (1,2) , $Q \in \text{Ext}^{3,7}$ <u>and</u> $P \in \text{Ext}^{4,12}$. (See e.g. [9, §6.1] for a proof.)

We mention in passing that Q has Massey product representation $\langle h_1^2,h_1,h_0\rangle$. Also, P has representation $\langle h_1,h_1^2,h_1,h_1^2\rangle$.

11.2 and 11.3 together determine the E^1 term of our spectral sequence. We now turn to the evaluation of the differentials. I know of no way to do this without going to the chain level. First, we need some notation. In E^1 , we denote an element in the kth copy of $\text{Ext}_{G_1 \otimes E(Q_3\cdots)}(Z_2,Z_2)$ with a (k) on the right. Thus $\theta \otimes q_3^{i_1}\cdots q_r^{i_r}$ in the kth copy will be written

$$\theta \otimes q_3^{i_1}\cdots q_r^{i_r}(k) .$$

The generators of C^0 are thus the $\{(k)_*\}$, and those in C^1 are the $h_0(k)_*$, $h_1(k)_*$, $q_3(k)_*$ $\cdots$. In principal, we can completely describe the differentials once we have specified the boundaries of the generators in $C^{(1)}$.

In the associated graded gadget, $\partial h_0(k) = Sq^1(k)_*$, $\partial h_1(k)_* = Sq^2(k)_*$, $\partial q_1(k) = Q_1(k)_*$. Of course, these are not in the kernel of the surjection $C^0 \to H^*(G_L)/A$ in general, so there are higher filtration correction terms which must be added to these (presumptive) boundaries. In particular,

$$\text{(11.4)} \qquad \begin{aligned} &Sq^2(k)_* + \binom{k-1}{2}(k+2)_* , \\ &Sq^1(k)_* + \binom{k-1}{1}(k+1)_* \end{aligned}$$

are in the kernel, and hence are appropriate choices for the actual differentials. Now, using standard calculational techniques, we have

$$\text{(11.5)} \qquad Q_1 Sq^k = Sq^{(0\cdots 02)}Sq^{k+1} + Sq^{(0\cdots 04)}Sq^{k+3} + \theta$$

where θ vanishes on elements of dimension $k+4$. Hence if we are concerned only with the first four or five differentials, we can assume

$$\text{(11.6)} \qquad \partial q_1(k)_* = Q_1(k)_* + Sq^{(0\cdots 02)}(k+1)_* + Sq^{(0\cdots 04)}(k+3)_* .$$

Thus we obtain

<u>Lemma</u> 11.7

$$\begin{aligned} \delta^1(k) &= \binom{k}{1}h_0(k-1) , \\ \delta^1(q_3(k)) &= \binom{k-1}{1}h_0 q_3(k-1) , \\ \delta^2(k) &= \binom{k-3}{2}h_1(k-2) , \\ \delta^2(q_3(k)) &= \binom{k-2}{2}h_1 q_3(k-2) . \end{aligned}$$

These differentials and some routine algebra allow the complete determination of $E^4_{s,r,t}$ in dimensions $r-s < 12$. For example, when $n = 4L+1$, we find

(11.8)

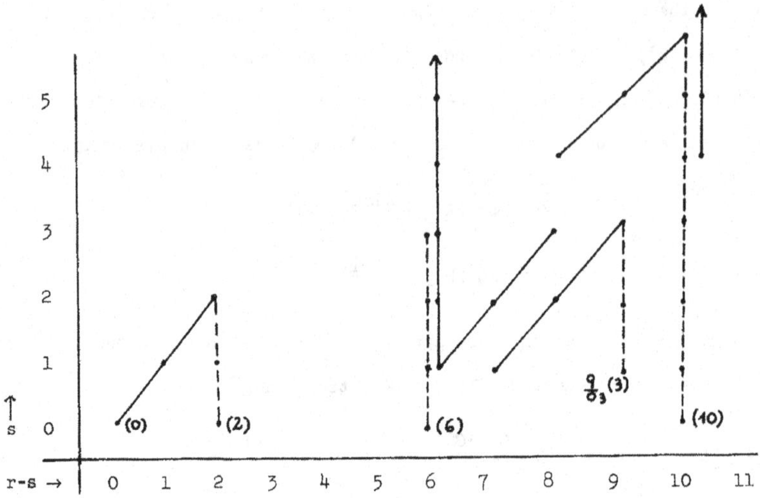

Our conventions for writing the Ext groups are the usual ones (e.g., see [16]). However, for the reader's convenience, we list them here. A vertical line connecting two dots represents the fact that the upper is h_0 times the lower. A 45° line connecting two dots says that the upper is h_1 times the lower. Finally, a dashed vertical line implies again that the upper is h_0 times the lower, but the extension is not obvious in E^1. (Actually, the precise significance is δ (lower) = $(Sq^1)^*$ (upper) + θ where θ has filtration at least 4 higher.)

Examining (11.8), we see that the only possible differential is

$\delta^7(10) = \begin{cases} q_3(3) \\ 0 \end{cases}$. In fact, we have

<u>Lemma</u> 11.9 $\delta^7(10) = \begin{cases} q_3(3), & n \equiv 1(8) \\ 0, & n \equiv 5(8) \end{cases}$. (For the proof, it is necessary only to expand (11.6) two further stages in these cases.)

Thus, in view of the dashed vertical lines, this determines the remaining differentials, and (11.8) represents $\mathrm{Ext}_{G(2)}(H^*(\Gamma_n)/A, Z_2)$ for $n \equiv 5(8)$, while

$$\mathrm{Ext}_{G(2)}(H^*(\Gamma_n)/A, Z_2)$$

for $n \equiv 1(8)$ has the form

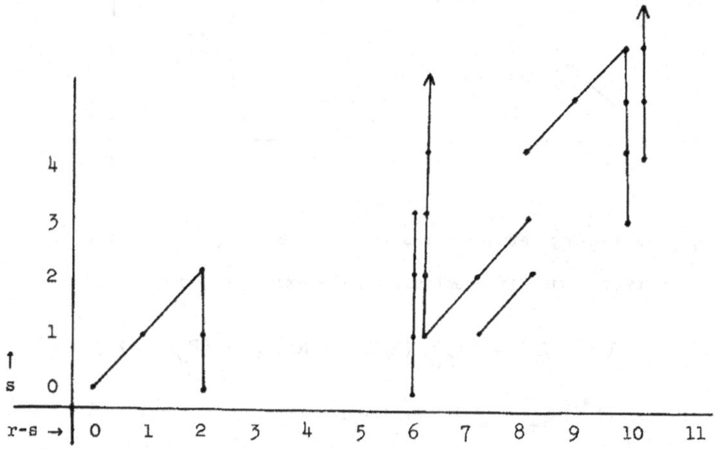

Now, following our program, we turn to the calculation of $\mathrm{Ext}_{G(2)}(A,Z_2)$ Unfortunately, I know no indirect way of studying this group, so I found it necessary to write down an explicit resolution in our range of interest. We suppress the calculations (in the interest of mercy to all concerned) and record the result:

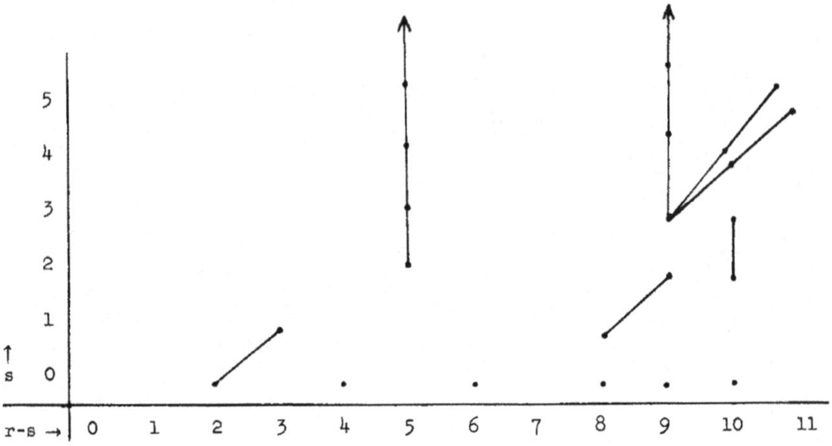

with, perhaps, some missing terms in $r-s = 11$. Combining this with (11.10), and calculating some obvious differentials in the exact sequence

$$(11.12) \quad \ldots \mathrm{Ext}_{G(2)}(H^*(F_L), Z_2) \to \mathrm{Ext}_{G(2)}(A,Z_2) \xrightarrow{\delta} \mathrm{Ext}_{G(2)}(H^*(F_L)/A, Z_2) \to \ldots ,$$

we find that $\mathrm{Ext}_{G(2)}(H^*(F_{8k+1}), Z_2)$ has the form

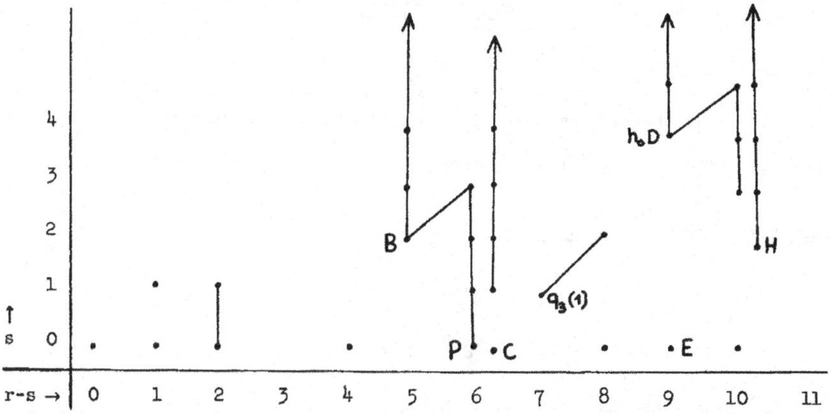

There are a number of "suprising" facts concealed in (11.13). For one, note that δ in (11.12) is highly non-trivial. For another, note e.g. that, for h_0D , H , there are extensions of the $Ext_{G(2)}(Z_2,Z_2)$ module structure connecting the remaining terms from (11.10) and (11.11). The original interpretation of elements in $Ext_{G(2)}(H^*(F_n)/A, Z_2)$ was as "stable" higher order cohomology operations which vanish universally on integral cohomology classes of dimension n (e.g. see [11]). However, it is clear that that hypothesis can no longer be supported since the terms from $Ext_{G(2)}(A,Z_2)$ play a very definite role in determining the actual Ext groups of F_n , and the differential depends on n . Thus the general defining system for an operation represented in $Ext_{G(2)}(H^*(F_n)/A, Z_2)$ may well have been built on relations which depend on lower order elements in $Ext_{G(2)}(A,Z_2)$ and n .

We now calculate the differentials in (11.13), and conclude this section by evaluating

$$E^2_{**}(F_n, A, Z_2)$$

in this range.

We first observe that the Z_4 Bockstein (Proposition 3.6) implies some ∂_2 differentials. Thus we have

<u>Lemma</u> 11.14 <u>In</u> (11.13), <u>we have</u>

$$\partial_2 g_3(0) = h_o B ,$$
$$\partial_2(H) = h_o D .$$

Somewhat less immediately, we have

<u>Lemma</u> 11.15 (i) $\partial_2(g_3(1)) = 0$,

(ii) $\partial_2(E) = 0$.

<u>Proof</u> (i) Consider the inclusion

$$u : S^1 \ltimes_T K \wedge K \to S^{n-1} \ltimes_T K \wedge K .$$

It is easy to verify that $g_3(1)$ is present in im (u_*) on Ext groups. However, $h_o{}^3D$ is not in this image. Hence $\partial_2(g_3(1)) = u_*\partial_2\langle g_3(1)\rangle = 0$, and (i) follows. To prove (ii), note that E^* has representative cycle $\langle \dot{S}q^3, Sq^4Sq^2\rangle$. Consider the inclusion

$$I : K \wedge K \to S^{n-1} \ltimes_T K \wedge K .$$

$K \wedge K$ is a wedge of $K(Z_2,n)$'s and $K(Z,n)$'s in low dimensions. A direct calculation shows that $Sq^4Sq^2\iota \otimes Sq^3\iota$ is a generator in $H^*(K \wedge K)$ over

$G(2)$. Hence it is dual to an element b in $\pi_*(K \wedge K)$. Clearly, $I_*(b)$ is detected by E^* , and (ii) follows.

Remark 11.16 The splitting of $K \wedge K$ used in the proof of 11.15(ii), together with the cofibrations

$$(11.17)\quad S^r \ltimes_T K \wedge K \to S^{r+1} \ltimes_T K \wedge K \to \Sigma^{r+1} K \wedge K \to \Sigma S^r \ltimes_T K \wedge K \to \dots ,$$

provide us with a good method for constructing homotopy classes in our range, while the global calculations of Ext groups provide an effective way to limit the number of elements whose study is required. It would seem to the author that this remark provides the reader with effective tools for analyzing the stable homotopy of Eilenberg-MacLane spaces so far as desired in our range!

Returning to our calculation, we easily see that 11.14 and 11.15 give all the ∂_2 differentials. Moreover, $E^3 \cong E^\infty$, and we have

Theorem 11.18 The first ten stable homotopy groups of $K(Z,8k+1)$ are given by the table

j	0	1	2	3	4	5	6	7	8	9
$\pi_{16k+j+1}$	Z_2	0	Z_4	0	Z_2	Z_2	$Z_2 \oplus Z_{16}$	Z_2	$Z_2 \oplus Z_2$	Z_2

.

For $j = 10$, $\pi^s_{16k+11}(K(Z,8k+1))$ has order at most 8 .

Remark 11.19 The groups obtained in 11.18 for $j = 6$, 7 do not agree with those obtained by Mahowald and Williams in [14]. They apparently missed the generator C corresponding to the cycle $\langle Sq^2, Sq^4\rangle$.

Finally, we apply 11.18 to calculate the differential $\partial : \mathrm{Ext}^{i}_{G(2)}(H^*(F_n)/A, Z_2) \to \mathrm{Ext}^{i+2}_{G(2)}(A, Z_2)$ defined in 6.11. We easily find that

$$E^2_*(F_{8k+1}, A, Z_2)$$

has the form

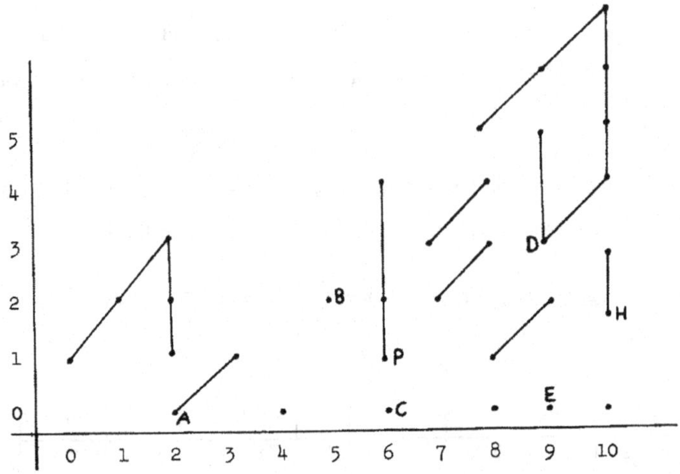

Again using 11.18, the reader can easily calculate the higher differentials in (11.20).

12. An Example For The Metastable Exact Sequence

In [13], Mahowald used his special case of 8.5 to give extensive calculations of the metastable homotopy groups of spheres. The most elementary spaces after the spheres are the various two-cell complexes; in particular, the four such non-trivial ones distinguished by their structure as $\mathcal{Q}(2)$-modules.

In this section, we apply 8.5 to calculate some of the unstable groups of the space $X_7 = S^7 \cup_2 e^8$ as an example of the techniques involved.

The stable homotopy of X_7 is easily studied. We have an exact sequence

(12.1) $$\xrightarrow{\partial} \mathrm{Ext}_{\mathcal{Q}(2)}(Z_2, Z_2) \xrightarrow{I} \mathrm{Ext}_{\mathcal{Q}(2)}(H^*(X_7), Z_2) \xrightarrow{J} \mathrm{Ext}_{\mathcal{Q}(2)}(Z_2, Z_2) \rightarrow ,$$

and ∂ is exactly multiplication by h_0, $\partial x = h_0 X$ (eg., see [2]). Hence $\mathrm{Ext}_{\mathcal{Q}(2)}(H^*(X_7), Z_2)$ consists of elements of two kinds:

(12.2) i) elements α in the image of I,

ii) elements β for which $J(\beta) \neq 0$ and $h_0 J(\beta) = 0$.

Elements of this second kind may be written explicitly as Massey products in $\mathrm{Ext}_{\mathcal{Q}(2)}(H^*(X_7), Z_2)$. Exactly, we have

(12.3) $$\beta = \langle J(\beta), h_0, \underset{\sim}{\iota} \rangle$$

where $\underset{\sim}{\iota} = I(\iota)$ is the unique non-zero element in $\mathrm{Ext}^{0,7}_{\mathcal{Q}(2)}(H^*(X_7), Z_2)$.
Consequently, we obtain

<u>Proposition</u> 12.4 <u>The following extensions occur in</u> $\mathrm{Ext}_{G(2)}(H^*(X_7), Z_2)$:

i) $h_0\langle \gamma, h_0 \underset{\sim}{\iota}\rangle = h_1 I(\gamma)\underset{\sim}{\iota}$;

ii) <u>If</u> $\alpha\gamma = 0$ <u>in</u> $\mathrm{Ext}_{G(2)}(Z_2, Z_2)$, <u>then</u> $\alpha\langle \gamma, h_0, \underset{\sim}{\iota}\rangle = \langle \alpha, \gamma, h_0\rangle \underset{\sim}{\iota}$.

<u>Proof</u> By the slide formula for Massey products,

$$\alpha\langle \beta, \gamma, \delta\rangle = \langle \alpha, \beta, \gamma\rangle \delta ,$$

we obtain (ii). To show (i), note that

$$h_0\langle \gamma, h_0, \underset{\sim}{\iota}\rangle = \langle h_0, \gamma, h_0\rangle \underset{\sim}{\iota}$$

from 12.2(ii) and 12.4(ii). On the other hand, the "Hirsch" formula ([17], [24]) $\langle a,b,a\rangle = (a \cup_1 a)b$ implies $\langle h_0, \gamma, h_0\rangle = h_1\gamma$ since $h_0 \cup_1 h_0 = h_1$, and (i) now follows.

Using the tabulated results ([16], [33]) on $\mathrm{Ext}_{G(2)}(Z_2, Z_2)$, we obtain $\mathrm{Ext}_{G(2)}(H^*(X_7), Z_2)$ for $t-s < 20$ as

$L = \langle h_1, h_0, \underset{\sim}{\iota} \rangle$

$M = \langle h_2^2, h_0, \underset{\approx}{\iota} \rangle$

$P = \langle h_0^3 h_3, h_0, \underset{\sim}{\iota} \rangle$

$\tilde{P} = \langle h_0, h_2^2, P \rangle$

$Q = \langle h_3^2, h_0^2, \underset{\approx}{\iota} \rangle$

$T = \langle h_0 f, h_0, \underset{\approx}{\iota} \rangle$

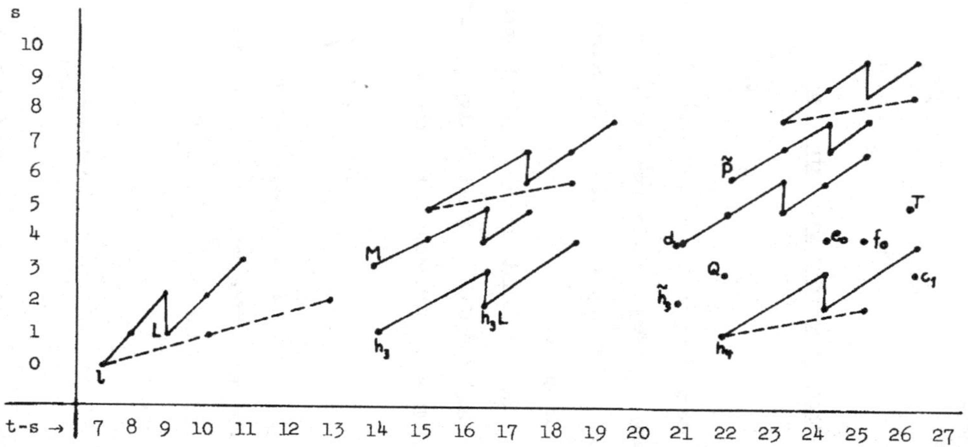

<u>Proposition</u> 12.6 <u>In the range of</u> (12.5), <u>there are only two non-zero differentials:</u>

$$\partial_2(e_0) = h_1^2 d_0 ,$$
$$\partial_2 T = h_1^2 d_0 L .$$

<u>Moreover, there is a non-trivial extension</u>

$$2\{h_4\} = Q .$$

<u>Proof</u> These differentials are non-trivial since the sequence (12.1) is actually a long exact sequence of Adams spectral sequences, being induced from the obvious cofiber map of spaces, and the corresponding differentials are non-zero for spheres.

The next possibility for a non-zero differential is $h_4 L$. In fact, $\partial_2(h_4 L) = (2h_3^2)L = \{2\sigma^2\}L$ by naturality, but $2\sigma^2 L = \sigma^2\eta^2 \underset{\sim}{\iota} = 0$ since $\sigma^2\eta = 0$. Thus $h_4 L$ represents the Toda bracket

$$\left\{\sigma^2, (2,\eta^2), \begin{pmatrix} L \\ \underset{\sim}{\iota} \end{pmatrix}\right\}$$

in $\pi^S_*(X_7)$, and is indeed an infinite cycle.

There remains the possiblity of a differential on f . In the Adams spectral sequence for the sphere, $\partial_2(f_0) = 2h_2(d_0)$. Hence f_0 represents the Toda bracket

$$\{\upsilon(d_0), 2, \underset{\sim}{\iota}\}$$

in $\pi^S_*(X_7)$. This completes the first part of 12.6.

For the second assertion, note that h_4 is a permanent cycle representing the Toda bracket

$$\{\sigma^2,\ 2,\ \underset{\sim}{\iota}\}\ .$$

Hence

$$(12.7)\qquad 2\{\sigma^2,2,\underset{\sim}{\iota}\} = \{\sigma^2,2,\underset{\sim}{\iota}\}2 = \{\sigma^2,2,2\underset{\sim}{\iota}\} = \{\sigma^2,4,\underset{\sim}{\iota}\}\ .$$

But this last represents $\langle h_3{}^2, h_0{}^2, \underset{\sim}{\iota}\rangle = 0$, and 12.6 follows.

Remark 12.8 The reader might wonder why we did not use the slide formula in (12.7), obtaining $2\{\sigma^2, 2, \underset{\sim}{\iota}\} \subset \{2\sigma^2, 2\}\underset{\sim}{\iota} = \eta\sigma^2\underset{\sim}{\iota} = 0$. But these equations are valid only modulo the total indeterminacy which in this case is $2\pi^S_{22}(X_7)$.

Now we turn to the calculation of $\pi_*(S^7 \cup_2 e^8)$ in the metastable range. The fiber in the map

$$F_L \to S^7 \cup_2 e^8 \to \Omega^L(S^{L+7} \cup_2 e^{L+8})$$

is given as $\Omega S^{L-1} \ltimes_T [(S^7 \cup_2 e^8) \wedge (S^7 \cup_2 e^8)]$. The calculation of $\mathrm{Ext}^{**}_{\mathfrak{a}(2)}(H^*(F_L), A, Z_2)$ is routine using the first spectral sequence in §9 (in this case an exact sequence), and we find $\mathrm{Ext}^{s,r}_{\mathfrak{a}(2)}(H^*(F_L), A, Z_2)$ has the form

(12.9)

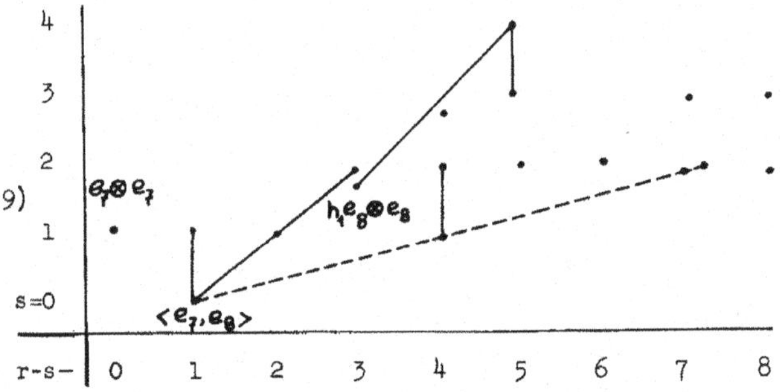

In particular, the reader should note that, since $\beta_4\langle e_7, e_8\rangle = \langle e_8 \otimes e_8\rangle$, it follows that the element γ, in the space $E_1(F_L, A)$, corresponding to $Sq^1(\iota_0)$ satisfies $Sq^1(\gamma) = e_8 \otimes e_8$. This is an example of the type of twisting referred to in §6, particularly 6.11. It is easy to check that there are no differentials in this range in the resulting Adams sequence, so $E^2 = E^\infty$.

We now evaluate the map ∂^1_2 of 8.5.

<u>Proposition</u> 12.10 $\partial_2(h_3\iota) = I_o$, <u>and there are no further non-trivial images for</u> $r-s < 8$.

<u>Proof</u> Since the k-invariant of $h_3\iota$ is $Sq^8(\iota_7)$, it is evident that $\partial_2(h_3\iota) = I_o$. The only other possibility is $\partial(h_3L) = \varepsilon h_1\langle e_7, e_8\rangle$, but this is part of the tower associated purely to A . Hence $\varepsilon = 0$, and 12.10 follows.

<u>Corollary</u> 12.11 <u>The unstable resolution of</u> $S^7 \cup_2 e^8$ <u>in the metastable range has the form</u>

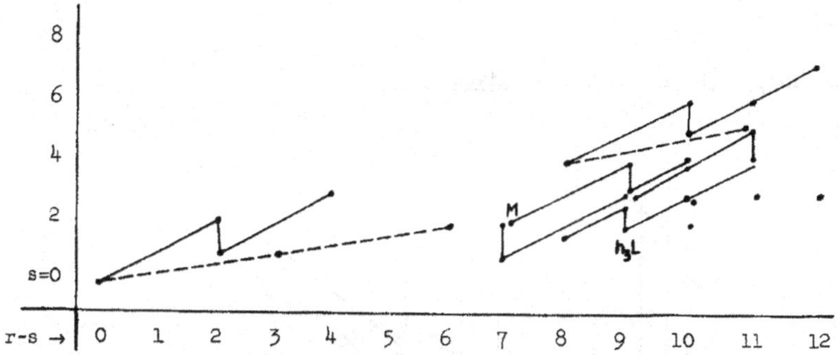

There are clearly no further differentials, so $E^2 = E^\infty$ in this range.

13. Further Calculations For Some Truncated Projective Spaces

In this section, we study the groups

$$(13.1)\qquad \begin{array}{ll} \pi_{i+n}(\Sigma^n(P^6_2)) , & i \leq 6 , \\ \pi_{i+n}(\Sigma^n(P^6)) , & i \leq 6 , n = 2 . \end{array}$$

They are of importance in the joint work of the author and E. Rees on embedding projective spaces ([42]), and provide examples illustrating much of the theory of Part I not already explored in §§9-12.

We being by recording the stable homotopy of the spaces in (13.1) and in our range.

__Lemma__ 13.2 $Ext_{G(2)}(H^*(P^6_2), Z_2)$ __is given by the table__

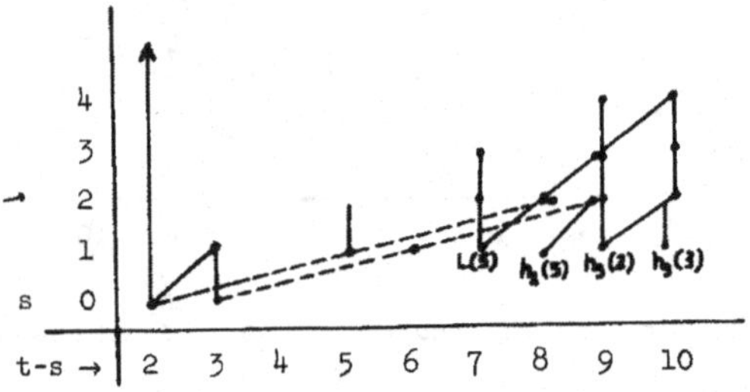

__in our range__.

Lemma 13.3 $\mathrm{Ext}_{G(2)}(H^*(P^6_2), Z_2)$ is given by the table

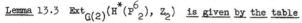

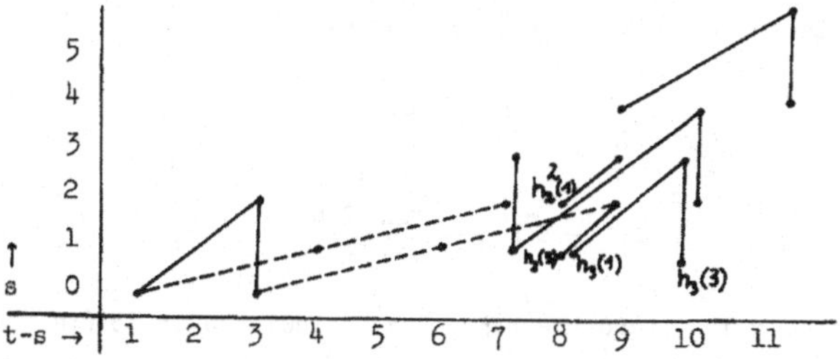

in our range. (Both 13.2 and 13.3 are evaluated by using the exact sequence of cofiberings

$$(13.4)\qquad \begin{array}{l} P_2^6 \xrightarrow{j} P_2^\infty \to P_7^\infty , \\ P_1^6 \to P_1^\infty \to P_7^\infty , \end{array}$$

taking the resulting long exact sequences of Ext groups, and using the tables in [13], pp. 54, 55, and 60 to evaluate $\mathrm{Ext}_{G(2)}(H^*(P_i^\infty), Z_2)$ in our range.)

From 13.2, 13.3, and the fact that, in $\pi_8(P_2^\infty)$, the class corresponding to $4L(5)$ is non-zero ([13], p. 55, and the observation $j_*(L(5)) = 2\{e_5\}$), it follows that $\partial_2 h_2(5) = 0$ in the Adams spectral sequences for which 13.2 and 13.3 are the E^2 terms. On the other hand, by inspection we see that $\partial_2(h_2(5))$ is the only possible non-zero differential. Thus $E^2 = E^\infty$ in both cases, and we have the table (localized at 2)

(13.5)

j	1	2	3	4	5	6	7	8	9
$\pi^s_j(P^6_2)$	0	Z	Z_4	0	Z_4	Z_2	Z_8	$(Z_2)^3$	$Z_{16} \oplus (Z_2)^2$
$\pi^s_j(P^6_1)$	Z_2	Z_2	Z_8	Z_2		Z_2	$Z_2 \oplus Z_8$	$(Z_2)^3$	$(Z_2)^5$.

Now we consider the unstable homotopy groups (13.1) for $\Sigma^2 P^6_1$, $\Sigma^2 P^6_2$. Since these are suspensions, the fibers

(13.6)
$$\begin{array}{ccccc} F_L^{\,1} & \to & \Sigma^2 P^6_1 & \to & \Omega^L \Sigma^{L+2} P^6_1 \\ \downarrow & & & & \downarrow \\ F_L^{\,2} & \to & \Sigma^2 P^6_2 & \to & \Omega^L \Sigma^{L+2} P^6_2 \end{array}$$

are given explicitly in §1 and their homology calculated in §3 (particularly 3.9, 3.10). There is no difficulty in constructing resolutions to evaluate $\pi_*(F_L^{\,\varepsilon})$ for $* \leq 8$. (For the fiber $F_L^{\,1}$, we are one dimension beyond the point where 1.11 is valid; thus the fact that we are dealing with suspensions is essential for effective calculation.)

Lemma 13.7 $\pi_7(F_L^{\,2}) = Z$ generator A,

$\pi_8(F_L^{\,2}) = Z_4 \oplus Z_2$ generators B, C

with relation $\eta A = 2B$.

Proof We are in the range in which 1.11 applies. The fiber $F_L^{\,2}$ has homology generators

(13.8)

7	8	9
$\sigma(e_4 * e_4)$	$\sigma Q_1(e_4)$	$\sigma Q_2(e_4)$
	$\sigma(e_4 * e_5)$	$\sigma(e_5 * e_5)$
		$\sigma(e_4 * e_6)$

Passing to cohomology over $G(2)$, we apply 3.7 to show that

$$Sq^1(\sigma Q_1(e_4)^*) = \sigma(Q_2(e_4))^* ,$$

$$Sq^1(\sigma(e_4 * e_5))^* = Sq^2(\sigma(e_4 * e_4)) ,$$

and there are no further non-zero operations in (13.8). 13.7 now follows on taking $Ext_{\mathcal{Q}(2)}(H^*(F_L^{\ 2}), Z_2)$ in this range.

The situation for $F_L^{\ 1}$ is similar. However, we must also take into account the classes $\Sigma^2 e_1\langle\Sigma^2 e_1, \Sigma^2 e_2\rangle$, $\Sigma^2 e_2\langle\Sigma^2 e_1, \Sigma^2 e_2\rangle$ (in the notation of 3.9).

<u>Lemma</u> 13.8 $\pi_5(F_L^{\ 1}) = Z_2$ <u>generator</u> $D = \langle\Sigma^2 e_1 * \Sigma^2 e_1\rangle$

$\pi_6(F_L^{\ 1}) = Z_4$ <u>generator</u> $E = Q_1(\Sigma^2 e_1) + \langle\Sigma^2 e_1, \Sigma^2 e_2\rangle$

$\pi_7(F_L^{\ 1}) = Z_2$ <u>generator</u> $F = \langle\Sigma^2 e_1, \Sigma^2 e_3\rangle$

$\pi_8(F_L^{\ 1}) = Z_2^{(3)}$ <u>generators</u> $\begin{cases} G = \Sigma^2 e_1 \cdot \langle\Sigma^2 e_1, \Sigma^2 e_2\rangle \\ \upsilon D \\ H = \{\eta, 4, D\} \end{cases}$.

<u>Also, in the map</u> $\varphi : F_L^{\ 1} \to F_L^{\ 2}$, <u>we find</u> $\varphi_*(D) = \varphi_*(E) = \varphi^*{}_*(F) = \varphi_*(G) = 0$, <u>but</u>

$$\varphi_*(H) = \eta A .$$

<u>Proof</u> Through dimension 8 , $\pi_*(F_L^{\ 1}) = \pi_*^{\ s}(F_L^{\ 1})$ since $F_L^{\ 1}$ is 4-connected. We thus take a stable resolution. $H^*(F_L^{\ 1})$ has generators

(13.9)

5	6	7	8	9
$(e_3 * e_3)^*$	$Q_1(e_3)^*$	$Q_2(e_3)^*$	$Q_3(e_3)^*$	$Q_4(e_3)^*$
	$\langle e_3,e_4\rangle^*$	$(e_4 * e_4)^*$	$Q_1(e_4)^*$	$Q_2(e_4)^*$
		$\langle e_3,e_5\rangle^*$	$\langle e_3,e_6\rangle^*$	$(e_5 * e_5)^*$
			$\langle e_4,e_5\rangle^*$	$\langle e_3,e_7\rangle^*$
			$e_3\langle e_3,e_4\rangle^*$	$e_4\langle e_3,e_4\rangle^*$
				$\langle e_4,e_6\rangle^*$

in dimensions less than or equal to 9 . A basis over $\mathfrak{a}(2)$ is easily seen to be $(e_3 * e_3)^*$, $Q_1(e_3)^*$, $(e_4 * e_4)^*$, $\langle e_3,e_5\rangle^*$, $e_3\langle e_3,e_4\rangle^*$, $Q_4(e_3)^*$, $(e_5 * e_5)^*$ in this range. Relations are $Sq^1(Q_1(e_3))^* = Sq^1(e_4 * e_4)^* = 0$,

(13.10)

$$Sq^4(e_3 * e_3)^* = Sq^3(Q_1(e_3))^* ,$$

$$(Sq^4+Sq^3Sq^1)(e_3 * e_3)^* = Sq^2(e_4 * e_4)^* .$$

Next, since $\beta_4(e_4 * e_4) = \langle e_3,e_4\rangle + Q_1(e_3)$, we see that, in the Adams spectral sequence for F_L^1 , $\partial_2(e_4 * e_4) = h_0^2(Q_1(e_3))$. Moreover, there are no further differentials in our range, and $E^3 = E^\infty$.

Finally, noting the fact that

$$\varphi^*(e_4' * e_4')^* = (e_4 * e_4)^* ,$$

we see that the filtration 2 class, due to the second relation in (13.10), maps to ηA in $\pi_*(F_L^2)$.

<u>Corollary</u> 13.11 *The E^2 term of our spectral sequence 6.4 for $\Sigma^2 P_2^6$ has the form*

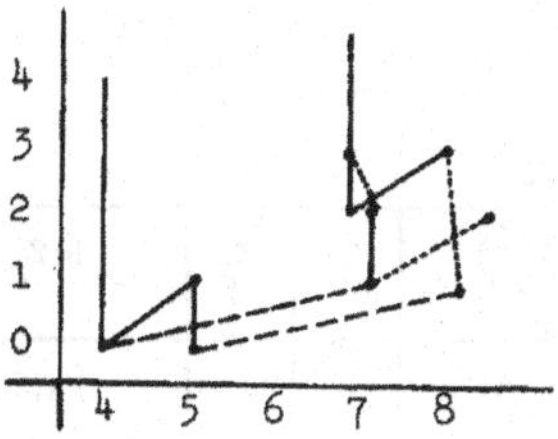

up to extensions (which are marked with dotted lines).

Remark 13.12 By using the techniques of §§1 and 8, we note that $j_*(G)$, $j_*(H)$ generate the unstable part of the homotopy group $\pi_8(\Sigma^2 P^6)$. Moreover, a simple argument with Whitehead products shows that $j_*(G)$ cannot be a suspension element. Thus $j_*(H)$ is the only candidate for an element in $\sigma^2\pi_6(P^6)$ in $\pi_8(\Sigma^2 P^6)$. In particular, the generator of $\pi_6(P^6)$ is the attaching map of the cell building P^7:

(13.13)
$$S^6 \xrightarrow{\lambda} P^6 \to P^7 \to S^7 ,$$

and it must be clear from (13.13) that the top class in $\Sigma^2 P^7$ is spherical if and only if $\sigma^2(\lambda) = 0$ in $\pi_8(\Sigma^2 P^6)$. But the only non-zero candidate is $j_*(H)$; and in $(\Sigma^2 P^6{}_2)$, $i_* j_*(H) \neq 0$ by 13.8. This proves the key result needed in [43]:

Theorem 13.13 *$\Sigma^2 P^7$ has top-class spherical if and only if $\Sigma^2 P^7{}_2$ does.*

We now complete the calculations of this paper by calculating $\pi_{2+i}(\Sigma P^6{}_2)$ and $\pi_{1+i}(P^6{}_2)$ in the range of (13.1).

Without difficulty, we find that the homotopy of the fiber $F_L{}^3$ in the map

$$\Sigma P^6{}_2 \to \Omega^L \Sigma^{L+1} P^6{}_2$$

is given by

(13.14)

j	5	6	7
$\pi_j(F_L^3)$	Z_2	Z_2	$Z_2 \oplus Z_2$
Generator	I	J	K, $\{\eta,2,I\}$.

Clearly, $\partial_2(\upsilon_2) = I$, and no other non-trivial boundary is possible. A schematic representation of $\pi_*(\Sigma P^6_2)$ in our range can now be given as

(13.15)

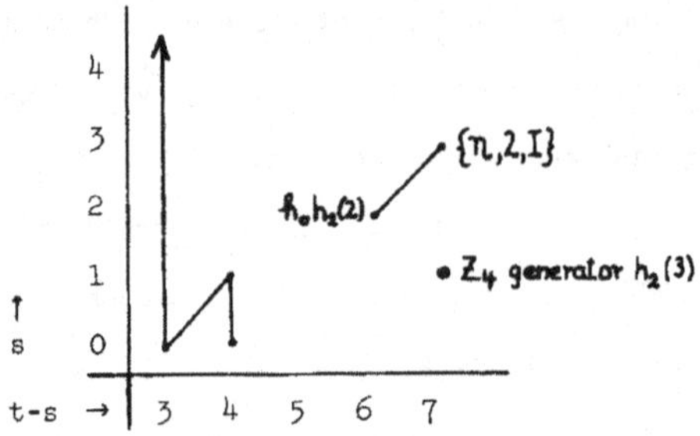

The extensions $2\upsilon(3) = j_*(K)$, $\eta[h_0h_2(2)] = j_*\{\eta,2,I\}$ must be verified. (Of course, the second extension follows directly from the fact that $\partial[h_2(2)] = I$.) To obtain the remaining extension, note that we are outside the range where 8.5 holds, so the extension is not necessarily surprising. (The point is, if 8.5 were true in this dimension, then in the resolution both K and $h_2(3)$ would occur in filtration 1 , and the extension $2\upsilon(3) = j_*K$ would clearly be impossible.)

Specifically, the difficulty with extending 8.5 occurs here because $Sq^4(\iota_4) = \iota_4^2$ is a Z_4 class, and $\beta_4(\iota_4)^2 = [Sq^1\iota_4 \cup \iota_4 + Sq^4Sq^1\iota_4]$ in $K(Z_2,4)$. Using the fact that we can take an unstable "resolution" of ΣP^6_2 with first K-invariants given as

$$(13.16)\quad \Sigma P^6_2 \xrightarrow{(\sigma e_2, \sigma e_3)} K(Z,3) \times K(Z_2,4) \xrightarrow{Sq^1 \iota_4 + Sq^2 \iota_3,\ \iota_4^2} K(Z_2,5) \times K(Z_4,8) ,$$

we easily justify (13.15), and this provides a good example of why the range of dimensions in which 8.5 holds cannot be extended.

Remark 13.17 We note in passing that, in the map $\mu : F_L^3 \to \Omega F_L^1$, we have $\mu_*(K) = 0$, $\mu_*\{\eta,2,I\} = \eta A$; and in the map $\mu : \Sigma P^6_2 \to \Omega\Sigma^2 P^6_2$, we have $\mu_*(h_o h_2(2)) = j_*(A) + 2\upsilon(2)$.

Now we conclude the discussion by studying briefly the homotopy of P^6_2 itself. It is a routine calculation with the Serre spectral sequence to give the homology of the fiber in the map

$$P_2{}^6 \to \Omega^L \Sigma^L P^6_2$$

in dimensions less than 8 . The fact that there are non-trivial cup products in P^6_2 produces some minor unpleasantries such as: (i) the class which should have transgressed to $(e_3 \circ e_4)^*$ is identified with the class transgressing to $(e_2 \circ e_5)^*$, and (ii) in $H^7(F)$, the class which transgresses to $(e_2 \circ e_2 \circ e_2 \circ e_2)^*$ is a Z_8-Bockstein,

$$\beta_8\{\sigma^{-1}(\langle e_3, e_4\rangle^* + e_1 \cup e_3 \otimes e_3)\} .$$

This illustrates the way in which the results of §1 fail when X is not a suspension.

The homotopy of F is given by

(13.18)

j	3	4	5	6
$\pi_j(F)$	Z	$Z_4 \oplus Z_2$	$Z_2 \oplus Z_2$	$Z_4 \oplus Z_8$
Generator	$[\iota_2, \iota_2]$	$\bar{K}$, L	$\eta\bar{K}$, $[\iota_3, \iota_3]$	M, P

Further, we have relations $\eta^2\bar{K} = 2M$, $\eta[\iota_2,\iota_2] = 2\bar{K}$, and $\Sigma(\bar{K}) = I$, $\Sigma(M) = \{\eta,2,I\}$ while $\Sigma P = K$ on suspending, i.e., taking the map

$$F \to \Omega F_L^3 .$$

Clearly, $\partial\upsilon(2) = \bar{K} \oplus L$, $\partial\upsilon(3) = [\iota_3,\iota_3]$. We may also verify $\partial(L(5)) = 2P$. This is not trivial; it involves the construction of an unstable resolution of P^6_2 , and makes essential use of the fact ([29]) that $\Sigma^2P^7_2$ is reducible. Here is what the resolution looks like through dimension 6 :

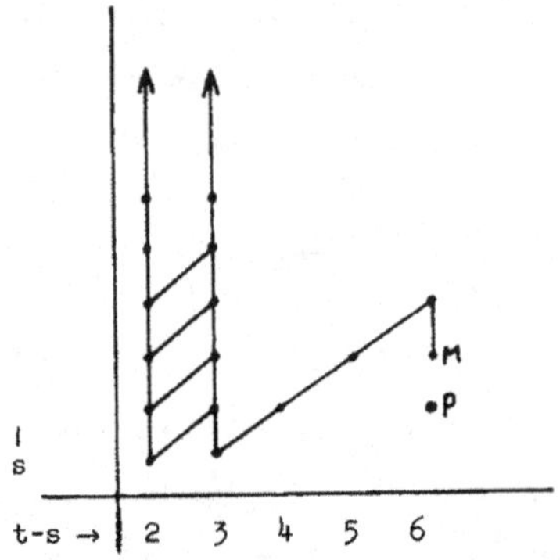

The k-invariants may be easily calculated. Notice that M has filtration 2 here but σM has filtration 3 in the resolution of ΣP^6_2 . This change of filtration degree would seem to merit further study (e.g., see [15] for an example of what can happen).

References

1. J. F. Adams, "On the non-existence of elements of Hopf invariant one," Ann. of Math. 72 (1960), 20-104.

2. ———, Stable Homotopy Theory, Lecture Notes in Mathematics 3, Springer-Verlag.

3. J. Adem, S. Gitler, "Secondary characteristic classes and the immersion problem," Bol. Soc. Mat. Mexicana 2 (1963), 53-78.

4. W. D. Barcus, "The stable suspension of an Eilenberg-MacLane space," Trans. Amer. Math. Soc. 96 (1960), 101-113.

5. W. D. Barcus, J. P. Meyer, "The suspension of a loop space," Amer. J. Math. 80 (1958), 895-920.

6. H. Cartan, "Sur les groupes d'Eilenberg-MacLane $H(\pi,n)$, I, II," Proc. Nat. Acad. Sci. U.S.A. 40 (1954), 464-471, 704-707.

7. M. Fuchs, "A modified Dold-Lashof construction that does classify H-principal fibrations," mimeo, Mich. State Univ., 1969.

8. S. Gitler, R. J. Milgram, "Unstable divisibility properties of the Chern character," to appear.

9. V. K. A. M. Gugenheim, R. J. Milgram, "On successive approximations in homological algebra," Trans. Amer. Math. Soc. 150 (1970), 157-182.

10. J. Harper, thesis, Univ. of Chicago, 1967.

11. ———, "Stable secondary cohomology operations," Comment. Math. Helv. 44 (1969), 341-353.

12. A. Liulevicius, "Multicomplexes and a general change of rings theorem," mimeo, Univ. of Chicago, 1967.

13. M. Mahowald, The Metastable Homotopy of S^n , Mem. Amer. Math. Soc. 72, 1967.

14. ———, R. Williams, "The stable homotopy of $K(Z,n)$," Bol. Soc. Mat. Mexicana 12 (1967), 22-28.

15. ———, "On the order of the image of J ," Topology 6 (1967), 371-378.

16. ———, M. Tangora, "Some differentials in the Adams spectral sequence," Topology 6 (1967), 349-369.

17. W. Massey, F. Peterson, The Mod 2 Cohomology of Certain Fiber Space, Mem. Amer. Math. Soc. 74, 1967.

18. R. J. Milgram, "Iterated loop spaces," Ann. of Math. 84 (1966), 386-403.

19. ———————, The bar construction and abelian H-spaces," Illinois J. Math. 11 (1967), 242-250.

20. ———————, "The homology of symmetric products," Trans. Amer. Math. Soc. 138 (1969), 251-265.

21. ———————, "The mod 2 spherical characteristic classes," Ann. of Math. 92 (1970), 238-261.

22. ———————, "Group representation and the Adams spectral sequence," to appear.

23. ———————, "The structure over the Steenrod algebra of some 2-stage Postnikov systems," Quart. J. Math. 20 (1969), 161-169.

24. ———————, "Steenrod squares and higher Massey products," Bol. Soc. Mat. Mexicana 13 (1968), 32-57.

25. J. Milnor, "On spaces having the homotopy type of a CW-complex," Trans. Amer. Math. Soc. 90 (1959), 272-280.

26. J. C. Moore, "Algèbre homologique et homologie des espaces classifiants," Sém. H. Cartan 12 (1959), 701-737.

27. M. Nakaoka, "Decomposition theorems for homology groups of symmetric groups," Ann. of Math. 71 (1960), 16-42.

28. T. Nishida, "Cohomology operations in iterated loop spaces," Proc. Japan Acad. 44 (1968), 104-109.

29. E. Rees, "Embedding theorems for real projective spaces," to appear.

30. L. Smith, J. C. Moore, "Hopf algebras and multiplicative fiber maps, I, II," Amer. J. Math. 90 (1968), 752-780, 1113-1150.

31. J. Stasheff, "Homotopy associativity of H-spaces II," Trans. Amer. Math. Soc. 108 (1963), 293-310.

32. N. Steenrod, "Milgram's classifying space for a topological group," Topology 7 (1968), 349-368.

33. M. Tangora, thesis, Northwestern Univ., 1966.

34. J. F. Adams, "Structure and applications of the Steenrod algebra," Comment. Math. Helv. 32 (1958), 180-214.

35. D. W. Anderson, "The e-invariant and the Hopf invariant," Topology 9 (1970), 49-54.

36. I. Berstein, T. Ganea, "Iterated suspensions," to appear, Comment. Math. Helv.

37. J. M. Boardman, R. M. Voigt, "Homotopy everything H-spaces," Bull. Amer. Math. Soc. 746 (1968), 1117-1122.

38. W. Browder, "The Kervaire invariant of framed manifolds and its generalizations," Ann. of Math. 90 (1969), 157-186.

39. E. H. Brown, "The Kervaire invariant of a manifold," Proc. Amer. Math. Soc. Symposia in Pure Math. 22 (1971), 65-72.

40. G. Brumfiel, I. Madsen, R. J. Milgram, "PL-characteristic classes and cobordism," Bull. Amer. Math. Soc. 77 (1971), 1025-1030.

41. R. J. Milgram, "Surgery with coefficients," mimeo, Stanford Univ., 1972.

42. ———————, E. Rees, "On the normal bundle to an embedding, to appear, Topology.

43. J. Milnor, J. C. Moore, "On the structure of Hopf algebras," Ann. of Math. 81 (1965), 211-264.

44. D. Quillen, "Rational homotopy theory," Ann. of Math. 90 (1969), 205-295.

Vol. 215: P. Antonelli, D. Burghelea and P. J. Kahn, The Concordance-Homotopy Groups of Geometric Automorphism Groups. X, 140 pages. 1971. DM 16,-

Vol. 216: H. Maaß, Siegel's Modular Forms and Dirichlet Series. VII, 328 pages. 1971. DM 20,-

Vol. 217: T. J. Jech, Lectures in Set Theory with Particular Emphasis on the Method of Forcing. V, 137 pages. 1971. DM 16,-

Vol. 218: C. P. Schnorr, Zufälligkeit und Wahrscheinlichkeit. IV, 212 Seiten. 1971. DM 20,-

Vol. 219: N. L. Alling and N. Greenleaf, Foundations of the Theory of Klein Surfaces. IX, 117 pages. 1971. DM 16,-

Vol. 220: W. A. Coppel, Disconjugacy. V, 148 pages. 1971. DM 16,-

Vol. 221: P. Gabriel und F. Ulmer, Lokal präsentierbare Kategorien. V, 200 Seiten. 1971. DM 18,-

Vol. 222: C. Meghea, Compactification des Espaces Harmoniques. III, 108 pages. 1971. DM 16,-

Vol. 223: U. Felgner, Models of ZF-Set Theory. VI, 173 pages. 1971. DM 16,-

Vol. 224: Revètements Etales et Groupe Fondamental. (SGA 1). Dirigé par A. Grothendieck XXII, 447 pages. 1971. DM 30,-

Vol. 225: Théorie des Intersections et Théorème de Riemann-Roch. (SGA 6). Dirigé par P. Berthelot, A. Grothendieck et L. Illusie. XII, 700 pages. 1971. DM 40,-

Vol. 226: Seminar on Potential Theory, II. Edited by H. Bauer. IV, 170 pages. 1971. DM 18,-

Vol. 227: H. L. Montgomery, Topics in Multiplicative Number Theory. IX, 178 pages. 1971. DM 18,-

Vol. 228: Conference on Applications of Numerical Analysis. Edited by J. Ll. Morris. X, 358 pages. 1971. DM 26,-

Vol. 229: J. Väisälä, Lectures on n-Dimensional Quasiconformal Mappings. XIV, 144 pages. 1971. DM 16,-

Vol. 230: L. Waelbroeck, Topological Vector Spaces and Algebras. VII, 158 pages. 1971. DM 16,-

Vol. 231: H. Reiter, L^1-Algebras and Segal Algebras. XI, 113 pages. 1971. DM 16,-

Vol. 232: T. H. Ganelius, Tauberian Remainder Theorems. VI, 75 pages. 1971. DM 16,-

Vol. 233: C. P. Tsokos and W. J. Padgett. Random Integral Equations with Applications to stochastic Systems. VII, 174 pages. 1971. DM 18,-

Vol. 234: A. Andreotti and W. Stoll. Analytic and Algebraic Dependence of Meromorphic Functions. III, 390 pages. 1971. DM 26,-

Vol. 235: Global Differentiable Dynamics. Edited by O. Hájek, A. J. Lohwater, and R. McCann. X, 140 pages. 1971. DM 16,-

Vol. 236: M. Barr, P. A. Grillet, and D. H. van Osdol. Exact Categories and Categories of Sheaves. VII, 239 pages. 1971. DM 20,-

Vol. 237: B. Stenström, Rings and Modules of Quotients. VII, 136 pages. 1971. DM 16,-

Vol. 238: Der kanonische Modul eines Cohen-Macaulay-Rings. Herausgegeben von Jürgen Herzog und Ernst Kunz. VI, 103 Seiten. 1971. DM 16,-

Vol. 239: L. Illusie, Complexe Cotangent et Déformations I. XV, 355 pages. 1971. DM 26,-

Vol. 240: A. Kerber, Representations of Permutation Groups I. VII, 192 pages. 1971. DM 18,-

Vol. 241: S. Kaneyuki, Homogeneous Bounded Domains and Siegel Domains. V, 89 pages. 1971. DM 16,-

Vol. 242: R. R. Coifman et G. Weiss, Analyse Harmonique Non-Commutative sur Certains Espaces. V, 160 pages. 1971. DM 16,-

Vol. 243: Japan-United States Seminar on Ordinary Differential and Functional Equations. Edited by M. Urabe. VIII, 332 pages. 1971. DM 26,-

Vol. 244: Séminaire Bourbaki - vol. 1970/71. Exposés 382-399. V, 356 pages. 1971. DM 26,-

Vol. 245: D. E. Cohen, Groups of Cohomological Dimension One. V, 99 pages. 1972. DM 16,-

Vol. 246: Lectures on Rings and Modules. Tulane University Ring and Operator Theory Year, 1970-1971. Volume I. X, 661 pages. 1972. DM 40,-

Vol. 247: Lectures on Operator Algebras. Tulane University Ring and Operator Theory Year, 1970-1971. Volume II. XI, 786 pages. 1972. DM 40,-

Vol. 248: Lectures on the Applications of Sheaves to Ring Theory. Tulane University Ring and Operator Theory Year, 1970-1971. Volume III. VIII, 315 pages. 1971. DM 26,-

Vol. 249: Symposium on Algebraic Topology. Edited by P. J. Hilton. VII, 111 pages. 1971. DM 16,-

Vol. 250: B. Jónsson, Topics in Universal Algebra. VI, 220 pages. 1972. DM 20,-

Vol. 251: The Theory of Arithmetic Functions. Edited by A. A. Gioia and D. L. Goldsmith VI, 287 pages. 1972. DM 24,-

Vol. 252: D. A. Stone, Stratified Polyhedra. IX, 193 pages. 1972. DM 18,-

Vol. 253: V. Komkov, Optimal Control Theory for the Damping of Vibrations of Simple Elastic Systems. V, 240 pages. 1972. DM 20,-

Vol. 254: C. U. Jensen, Les Foncteurs Dérivés de $\varprojlim$ et leurs Applications en Théorie des Modules. V, 103 pages. 1972. DM 16,-

Vol. 255: Conference in Mathematical Logic - London '70. Edited by W. Hodges. VIII, 351 pages. 1972. DM 26,-

Vol. 256: C. A. Berenstein and M. A. Dostal, Analytically Uniform Spaces and their Applications to Convolution Equations. VII, 130 pages. 1972. DM 16,-

Vol. 257: R. B. Holmes, A Course on Optimization and Best Approximation. VIII, 233 pages. 1972. DM 20,-

Vol. 258: Séminaire de Probabilités VI. Edited by P. A. Meyer. VI, 253 pages. 1972. DM 22,-

Vol. 259: N. Moulis, Structures de Fredholm sur les Variétés Hilbertiennes. V, 123 pages. 1972. DM 16,-

Vol. 260: R. Godement and H. Jacquet, Zeta Functions of Simple Algebras. IX, 188 pages. 1972. DM 18,-

Vol. 261: A. Guichardet, Symmetric Hilbert Spaces and Related Topics. V, 197 pages. 1972. DM 18,-

Vol. 262: H. G. Zimmer, Computational Problems, Methods, and Results in Algebraic Number Theory. V, 103 pages. 1972. DM 16,-

Vol. 263: T. Parthasarathy, Selection Theorems and their Applications. VII, 101 pages. 1972. DM 16,-

Vol. 264: W. Messing, The Crystals Associated to Barsotti-Tate Groups: With Applications to Abelian Schemes. III, 190 pages. 1972. DM 18,-

Vol. 265: N. Saavedra Rivano, Catégories Tannakiennes. II, 418 pages. 1972. DM 26,-

Vol. 266: Conference on Harmonic Analysis. Edited by D. Gulick and R. L. Lipsman. VI, 323 pages. 1972. DM 24,-

Vol. 267: Numerische Lösung nichtlinearer partieller Differential- und Integro-Differentialgleichungen. Herausgegeben von R. Ansorge und W. Törnig, VI, 339 Seiten. 1972. DM 26,-

Vol. 268: C. G. Simader, On Dirichlet's Boundary Value Problem. IV, 238 pages. 1972. DM 20,-

Vol. 269: Théorie des Topos et Cohomologie Etale des Schémas. (SGA 4). Dirigé par M. Artin, A. Grothendieck et J. L. Verdier. XIX, 525 pages. 1972. DM 50,-

Vol. 270: Théorie des Topos et Cohomologie Etale des Schémas. Tome 2. (SGA 4). Dirigé par M. Artin, A. Grothendieck et J. L. Verdier. V, 418 pages. 1972. DM 50,-

Vol. 271: J. P. May, The Geometry of Iterated Loop Spaces. IX, 175 pages. 1972. DM 18,-

Vol. 272: K. R. Parthasarathy and K. Schmidt, Positive Definite Kernels, Continuous Tensor Products, and Central Limit Theorems of Probability Theory. VI, 107 pages. 1972. DM 16,-

Vol. 273: U. Seip, Kompakt erzeugte Vektorräume und Analysis. IX, 119 Seiten. 1972. DM 16,-

Vol. 274: Toposes, Algebraic Geometry and Logic. Edited by. F. W. Lawvere. VI, 189 pages. 1972. DM 18,-

Vol. 275: Séminaire Pierre Lelong (Analyse) Année 1970-1971. VI, 181 pages. 1972. DM 18,-

Vol. 276: A. Borel, Représentations de Groupes Localement Compacts. V, 98 pages. 1972. DM 16,-

Vol. 277: Séminaire Banach. Edité par C. Houzel. VII, 229 pages. 1972. DM 20,-

Vol. 278: H. Jacquet, Automorphic Forms on GL(2). Part II. XIII, 142 pages. 1972. DM 16,–

Vol. 279: R. Bott, S. Gitler and I. M. James, Lectures on Algebraic and Differential Topology. V, 174 pages. 1972. DM 18,–

Vol. 280: Conference on the Theory of Ordinary and Partial Differential Equations. Edited by W. N. Everitt and B. D. Sleeman. XV, 367 pages. 1972. DM 26,–

Vol. 281: Coherence in Categories. Edited by S. Mac Lane. VII, 235 pages. 1972. DM 20,–

Vol. 282: W. Klingenberg und P. Flaschel, Riemannsche Hilbertmannigfaltigkeiten. Periodische Geodätische. VII, 211 Seiten. 1972. DM 20,–

Vol. 283: L. Illusie, Complexe Cotangent et Déformations II. VII, 304 pages. 1972. DM 24,–

Vol. 284: P. A. Meyer, Martingales and Stochastic Integrals I. VI, 89 pages. 1972. DM 16,–

Vol. 285: P. de la Harpe, Classical Banach-Lie Algebras and Banach-Lie Groups of Operators in Hilbert Space. III, 160 pages. 1972. DM 16,–

Vol. 286: S. Murakami, On Automorphisms of Siegel Domains. V, 95 pages. 1972. DM 16,–

Vol. 287: Hyperfunctions and Pseudo-Differential Equations. Edited by H. Komatsu. VII, 529 pages. 1973. DM 36,–

Vol. 288: Groupes de Monodromie en Géométrie Algébrique. (SGA 7 I). Dirigé par A. Grothendieck. IX, 523 pages. 1972. DM 50,–

Vol. 289: B. Fuglede, Finely Harmonic Functions. III, 188. 1972. DM 18,–

Vol. 290: D. B. Zagier, Equivariant Pontrjagin Classes and Applications to Orbit Spaces. IX, 130 pages. 1972. DM 16,–

Vol. 291: P. Orlik, Seifert Manifolds. VIII, 155 pages. 1972. DM 16,–

Vol. 292: W. D. Wallis, A. P. Street and J. S. Wallis, Combinatorics: Room Squares, Sum-Free Sets, Hadamard Matrices. V, 508 pages. 1972. DM 50,–

Vol. 293: R. A. DeVore, The Approximation of Continuous Functions by Positive Linear Operators. VIII, 289 pages. 1972. DM 24,–

Vol. 294: Stability of Stochastic Dynamical Systems. Edited by R. F. Curtain. IX, 332 pages. 1972. DM 26,–

Vol. 295: C. Dellacherie, Ensembles Analytiques, Capacités, Mesures de Hausdorff. XII, 123 pages. 1972. DM 16,–

Vol. 296: Probability and Information Theory II. Edited by M. Behara, K. Krickeberg and J. Wolfowitz. V, 223 pages. 1973. DM 20,–

Vol. 297: J. Garnett, Analytic Capacity and Measure. IV, 138 pages. 1972. DM 16,–

Vol. 298: Proceedings of the Second Conference on Compact Transformation Groups. Part 1. XIII, 453 pages. 1972. DM 32,–

Vol. 299: Proceedings of the Second Conference on Compact Transformation Groups. Part 2. XIV, 327 pages. 1972. DM 26,–

Vol. 300: P. Eymard, Moyennes Invariantes et Représentations Unitaires. II, 113 pages. 1972. DM 16,–

Vol. 301: F. Pittnauer, Vorlesungen über asymptotische Reihen. VI, 186 Seiten. 1972. DM 18,–

Vol. 302: M. Demazure, Lectures on p-Divisible Groups. V, 98 pages. 1972. DM 16,–

Vol. 303: Graph Theory and Applications. Edited by Y. Alavi, D. R. Lick and A. T. White. IX, 329 pages. 1972. DM 26,–

Vol. 304: A. K. Bousfield and D. M. Kan, Homotopy Limits, Completions and Localizations. V, 348 pages. 1972. DM 26,–

Vol. 305: Théorie des Topos et Cohomologie Etale des Schémas. Tome 3. (SGA 4). Dirigé par M. Artin, A. Grothendieck et J. L. Verdier. VI, 640 pages. 1973. DM 50,–

Vol. 306: H. Luckhardt, Extensional Gödel Functional Interpretation. VI, 161 pages. 1973. DM 18,–

Vol. 307: J. L. Bretagnolle, S. D. Chatterji et P.-A. Meyer, Ecole d'été de Probabilités: Processus Stochastiques. VI, 198 pages. 1973. DM 20,–

Vol. 308: D. Knutson, λ-Rings and the Representation Theory of the Symmetric Group. IV, 203 pages. 1973. DM 20,–

Vol. 309: D. H. Sattinger, Topics in Stability and Bifurcation Theory. VI, 190 pages. 1973. DM 18,–

Vol. 310: B. Iversen, Generic Local Structure of the Morphisms in Commutative Algebra. IV, 108 pages. 1973. DM 16,–

Vol. 311: Conference on Commutative Algebra. Edited by J. W. Brewer and E. A. Rutter. VII, 251 pages. 1973. DM 22,–

Vol. 312: Symposium on Ordinary Differential Equations. Edited by W. A. Harris, Jr. and Y. Sibuya. VIII, 204 pages. 1973. DM 22,–

Vol. 313: K. Jörgens and J. Weidmann, Spectral Properties of Hamiltonian Operators. III, 140 pages. 1973. DM 16,–

Vol. 314: M. Deuring, Lectures on the Theory of Algebraic Functions of One Variable. VI, 151 pages. 1973. DM 16,–

Vol. 315: K. Bichteler, Integration Theory (with Special Attention to Vector Measures). VI, 357 pages. 1973. DM 26,–

Vol. 316: Symposium on Non-Well-Posed Problems and Logarithmic Convexity. Edited by R. J. Knops. V, 176 pages. 1973. DM 18,–

Vol. 317: Séminaire Bourbaki – vol. 1971/72. Exposés 400–417. IV, 361 pages. 1973. DM 26,–

Vol. 318: Recent Advances in Topological Dynamics. Edited by A. Beck, VIII, 285 pages. 1973. DM 24,–

Vol. 319: Conference on Group Theory. Edited by R. W. Gatterdam and K. W. Weston. V, 188 pages. 1973. DM 18,–

Vol. 320: Modular Functions of One Variable I. Edited by W. Kuyk. V, 195 pages. 1973. DM 18,–

Vol. 321: Séminaire de Probabilités VII. Edité par P. A. Meyer. VI, 32 pages. 1973. DM 26,–

Vol. 322: Nonlinear Problems in the Physical Sciences and Biology. Edited by I. Stakgold, D. D. Joseph and D. H. Sattinger. VIII, 35 pages. 1973. DM 26,–

Vol. 323: J. L. Lions, Perturbations Singulières dans les Problème aux Limites et en Contrôle Optimal. XII, 645 pages. 1973. DM 42,

Vol. 324: K. Kreith, Oscillation Theory. VI, 109 pages. 1973. DM 16,

Vol. 325: Ch.-Ch. Chou, La Transformation de Fourier Complexe e L'Equation de Convolution. IX, 137 pages. 1973. DM 16,–

Vol. 326: A. Robert, Elliptic Curves. VIII, 264 pages. 1973. DM 22,

Vol. 327: E. Matlis, 1-Dimensional Cohen-Macaulay Rings. XII, 15 pages. 1973. DM 18,–

Vol. 328: J. R. Büchi and D. Siefkes, The Monadic Second Orde Theory of All Countable Ordinals. VI, 217 pages. 1973. DM 20,

Vol. 329: W. Trebels, Multipliers for (C, α)-Bounded Fourier Expa sions in Banach Spaces and Approximation Theory. VII, 103 page 1973. DM 16,–

Vol. 330: Proceedings of the Second Japan-USSR Symposium o Probability Theory. Edited by G. Maruyama and Yu. V. Prokhoro VI, 550 pages. 1973. DM 36,–

Vol. 331: Summer School on Topological Vector Spaces. Edited L. Waelbroeck. VI, 226 pages. 1973. DM 20,–

Vol. 332: Séminaire Pierre Lelong (Analyse) Année 1971-1972. V, 1 pages. 1973. DM 16,–

Vol. 333: Numerische, insbesondere approximationstheoretische B handlung von Funktionalgleichungen. Herausgegeben von R. A sorge und W. Törnig. VI, 296 Seiten. 1973. DM 24,–

Vol. 334: F. Schweiger, The Metrical Theory of Jacobi-Perron Alg rithm. V, 111 pages. 1973. DM 16,–

Vol. 335: H. Huck, R. Roitzsch, U. Simon, W. Vortisch, R. Walden, Wegner und W. Wendland, Beweismethoden der Differentialgeom trie im Großen. IX, 159 Seiten. 1973. DM 18,–

Vol. 336: L'Analyse Harmonique dans le Domaine Complexe. Edi par E. J. Akutowicz. VIII, 169 pages. 1973. DM 18,–

Vol. 337: Cambridge Summer School in Mathematical Logic. Edit by A. R. D. Mathias and H. Rogers. IX, 660 pages. 1973. DM 42

Vol: 338: J. Lindenstrauss and L. Tzafriri, Classical Banach Space IX, 243 pages. 1973. DM 22,–

Vol. 339: G. Kempf, F. Knudsen, D. Mumford and B. Saint-Don Toroidal Embeddings I. VIII, 209 pages. 1973. DM 20,–

Vol. 340: Groupes de Monodromie en Géométrie Algébriq (SGA 7 II). Par P. Deligne et N. Katz. X, 438 pages. 1973. DM 40

Vol. 341: Algebraic K-Theory I, Higher K-Theories. Edited H. Bass. XV, 335 pages. 1973. DM 26,–

Vol. 342: Algebraic K-Theory II, "Classical" Algebraic K-Theo and Connections with Arithmetic. Edited by H. Bass. XV, 5 pages. 1973. DM 36,–

ᵛol. 343: Algebraic K-Theory III, Hermitian K-Theory and Geo- ɲetric Applications. Edited by H. Bass. XV, 572 pages. 1973. ⁾M 38,–

ᵛol. 344: A. S. Troelstra (Editor), Metamathematical Investigation ᵢf Intuitionistic Arithmetic and Analysis. XVII, 485 pages. 1973. ⁾M 34,–

ᵛol. 345: Proceedings of a Conference on Operator Theory. :dited by P. A. Fillmore. VI, 228 pages. 1973. DM 20,–

ᵛol. 346: Fučík et al., Spectral Analysis of Nonlinear Operators. , 287 pages. 1973. DM 26,–

ᵛol. 347: J. M. Boardman and R. M. Vogt, Homotopy Invariant ılgebraic Structures on Topological Spaces. X, 257 pages. 973. DM 22,–

ᵒol. 348: A. M. Mathai and R. K. Saxena, Generalized Hyper- eometric Functions with Applications in Statistics and Physical iciences. VII, 314 pages. 1973. DM 26,–

ᵒol. 349: Modular Functions of One Variable II. Edited by W. Kuyk nd P. Deligne. V, 598 pages. 1973. DM 38,–

ᵒol. 350: Modular Functions of One Variable III. Edited by W. uyk and J.-P. Serre. V, 350 pages. 1973. DM 26,–

ᵒol. 351: H. Tachikawa, Quasi-Frobenius Rings and Generaliza- ons. XI, 172 pages. 1973. DM 18,–

ɔl. 352: J. D. Fay, Theta Functions on Riemann Surfaces. V, 37 pages. 1973. DM 16,–

ol. 353: Proceedings of the Conference on Orders, Group ings and Related Topics. Organized by J. S. Hsia, M. L. Madan nd T. G. Ralley. X, 224 pages. 1973. DM 20,–

ol. 354: K. J. Devlin, Aspects of Constructibility. XII, 240 pages. 73. DM 22,–

ol. 355: M. Sion, A Theory of Semigroup Valued Measures. , 140 pages. 1973. DM 16,–

ɔl. 356: W. L. J. van der Kallen, Infinitesimally Central-Exten- ons of Chevalley Groups. VII, 147 pages. 1973. DM 16,–

ɔl. 357: W. Borho, P. Gabriel und R. Rentschler, Primideale Einhüllenden auflösbarer Lie-Algebren. V, 182 Seiten. 1973. M 18,–

ɔl. 358: F. L. Williams, Tensor Products of Principal Series epresentations. VI, 132 pages. 1973. DM 16,–

ɔl. 359: U. Stammbach, Homology in Group Theory. VIII, 183 ages. 1973. DM 18,–

ɔl. 360: W. J. Padgett and R. L. Taylor, Laws of Large Numbers r Normed Linear Spaces and Certain Fréchet Spaces. VI, 11 pages. 1973. DM 16,–

ol. 361: J. W. Schutz, Foundations of Special Relativity: Kine- atic Axioms for Minkowski Space Time. XX, 314 pages. 1973. M 26,–

l. 362: Proceedings of the Conference on Numerical Solution Ordinary Differential Equations. Edited by D. Bettis. VIII, 0 pages. 1974. DM 34,–

l. 363: Conference on the Numerical Solution of Differential uations. Edited by G. A. Watson. IX, 221 pages. 1974. DM 20,–

l. 364: Proceedings on Infinite Dimensional Holomorphy. ited by T. L. Hayden and T. J. Suffridge. VII, 212 pages. 1974. A 20,–

l. 365: R. P. Gilbert, Constructive Methods for Elliptic Equa- ns. VII, 397 pages. 1974. DM 26,–

l. 366: R. Steinberg, Conjugacy Classes in Algebraic Groups otes by V. V. Deodhar). VI, 159 pages. 1974. DM 18,–

l. 367: K. Langmann und W. Lütgebohmert, Cousinvertei- gen und Fortsetzungsätze. VI, 151 Seiten. 1974. DM 16,–

l. 368: R. J. Milgram, Unstable Homotopy from the Stable int of View. V, 109 pages. 1974. DM 16,–